HYDROXYAPATITE - ADVANCES IN COMPOSITE NANOMATERIALS, BIOMEDICAL APPLICATIONS AND ITS TECHNOLOGICAL FACETS

Edited by **Jagannathan Thirumalai**

Hydroxyapatite - Advances in Composite Nanomaterials, Biomedical Applications and Its Technological Facets
http://dx.doi.org/10.5772/intechopen.68820
Edited by Jagannathan Thirumalai

Contributors

Jagannathan Thirumalai, Wan Sharuzi Wan Harun, Rahil Izzati Mohd Asri, Saiful Anwar Che Ghani, Zakri Ghazalli, Abu Bakar Sulong, Rodica-Mariana Ion, Aurora Anca Poinescu, Jose Calegaro, Feray Bakan, Florin Miculescu, Marian Miculescu, Andreea Maidaniuc, Aura-Cătălina Mocanu, Ștefan Ioan Voicu, Cătălina-Andreea Dascălu, Robert-Cătălin Ciocoiu, Benedetta De Caprariis, Angelo Chianese, Marco Stoller, Nicola Verdone, Joanna Kolmas, Katarzyna Szurkowska, Aleksandra Laskus, Eric M. Rivera-Muñoz, Rodrigo Velázquez-Castillo, Lauro Bucio-Galindo, Miguel Apátiga-Castro, Beatriz Millán-Malo, Rufino Nava, Alejandro Manzano-Ramírez, Rafael Huirache-Acuña, Susana Alonso-Sierra, J. Rafael Alanís-Gómez

Published by InTech
Janeza Trdine 9, 51000 Rijeka, Croatia

Publishing Process Manager Teo Kos
Technical Editor SPi Global
Cover InTech Design team

First published February, 2018
Printed in Croatia
Legal deposit, Croatia: National and University Library in Zagreb

Additional hard copies can be obtained from orders@intechopen.com

Hydroxyapatite - Advances in Composite Nanomaterials, Biomedical Applications and Its Technological Facets, Edited by Jagannathan Thirumalai
p. cm.
Print ISBN 978-953-51-3804-4
Online ISBN 978-953-51-3805-1

For EU product safety concerns:
IN TECH d.o.o., Prolaz Marije Krucifikse Kozulić 3, 51000 Rijeka, Croatia,
info@intechopen.com or visit our website at intechopen.com.

Meet the editor

Dr. J. Thirumalai received his PhD degree from Alagappa University, Karaikudi, in 2010. He was also awarded the Post-doctoral Fellowship from Pohang University of Science and Technology (POSTECH), Republic of Korea, in 2013. He worked as an assistant professor in physics, B.S. Abdur Rahman University, Chennai, India (2011 to 2016). Currently, he is working as an assistant professor of physics, SASTRA University, Kumbakonam (T.N.), India. His research interests focus on luminescence, self-assembled nanomaterials and thin-film optoelectronic devices. He has published more than 50 Scopus/ISI indexed papers and 8 book chapters and is a member in several national and international societies like RSC, OSA, etc. He has edited two books for InTechOpen. He served as a principal investigator for a funded project towards the application of luminescence-based thin-film optoelectronic devices, funded by the Science and Engineering Research Board (SERB), India. As an expert in optoelectronics and nanotechnology, he has been invited as external and internal examiner to MSc and PhD theses. He has been invited to talk in various forums, and he also actively reviews papers for international and national journals.

Contents

Preface

Hydroxyapatite (HAp) has great importance in the fields of biomaterials and biomedical science, as well as in modern-day nanoscience and bio-science and their technological applications. HAp is one of only a few resources that are categorized as bioactive, i.e., it will support bone ingrowth and osseointegration when used in orthopedic, dental, and maxillofacial applications. A significant characteristic of HAp is its stability when compared with other calcium phosphates. HAp is commonly used as a filler to replace amputated bone or as a coating to promote bone ingrowth into prosthetic implants, thus achieving faster healing and recovery. Furthermore, HAp is used in drug delivery applications for orthopedic and dental implants. Recently, HAp was involved in various advances in nanotechnology-based regenerative biomedical applications for the repair or improvement of damaged tissue function in various organ systems.

The development of nanotechnology has had a major impact on the science of materials. The production of nanomaterials has gained considerable attention for adsorption, catalysis, and optical applications, particularly when biomaterials are involved. Knowledge of Hap-based nanomaterials provides a platform for innovation. This has the tendency to progress life requirements rather than scientific pursuit. Owing to its similarity in size, crystallography, and chemical opus with human hard tissue, the nano-HAp shows fascinating potential as a biomaterial for use in prosthetic applications. Apatite is the mineral found in bone and teeth in the form of enamel. HAp has exceptional properties such as bioactivity, biocompatibility, non-toxicity, osteoconductivity, and a non-inflammatory nature. Furthermore, the nano-HAp bioceramic has astonishing applications.

The aim of this book is to present information on nano-HAp materials. The book focuses on synthesis, properties, future prospects, and potential biomedical applications. Established researchers in this field have contributed different chapters, which cover advanced areas of research and developments in the field of materials science. This bookis intended for researchers and students wanting to gain a deeper understanding of the basic concepts of HAp that are applicable to many related fields.

This book consists of nine chapters that have been categorized into two sections. Section one consists of four chapters on HAp preparation, properties, perspectives, and process. This section introduces the different synthesis techniques and reviews nano-HAp and other novel materials-based regenerative biomedical applications, synthesis of nano-HAp, production of HAp through the intensification process, growth of crystalline HAp nanofibers through preferential orientation, and 153Sm-HAp for the cost-effective treatment of hemophilic arthropathy in large and medium joints. Section two contains five chapters on the efficient optimization of HAp-based coating for biomedical implants, preparation of calcium phosphates from marine and land bio-resources for biomimetic applications, Hap-based composite materials for bone tissue infections, and the bio-medical review of gene delivery [*in-vivo* RNA (siRNA) and microRNA (miRNA)] applications.

Finally, I would like to thank all of the contributors for their great efforts in producing an outstanding work. I would like to express my sincere gratitude to Mr. Teo Kos, publishing process manager, for his support in the generation of this book.

Jagannathan Thirumalai
SASTRA Deemed University, India

Section 1

Hydroxyapatite: Synthesis, Properties, Perspectives and its Prospects

Chapter 1

Introductory Chapter: The Testament of Hydroxyapatite: New Prospects in Regenerative Medicinal Treatments

Jagannathan Thirumalai

Additional information is available at the end of the chapter

http://dx.doi.org/10.5772/intechopen.72767

1. A terse testament of hydroxyapatite

The term 'Hydroxyapatite (HAp)' is a naturally occurring mineral and chemically identical to the mineral constituent of bones and solid tissues of mankind and mammals. As a mineral species, apatite was first known in 1786 by "the father of German geology" Abraham Gottlob Werner (1750–1817) and entitled by him on or after the antediluvian Greek $\alpha\pi\alpha\tau\acute{\alpha}\omega$ (apatao)—"to mislead" or "to deceive" since it had earlier been does not specify one chemical opus. Though, the word "apatite" was revealed in the 1990s and is denoted as calcium orthophosphate, which would be a very infrequent heterogeneity of tourmaline, beryl and other stones [1]. The period of HAp in reformative science backdate to 1950s [2] furthermore for the filling of the bone defects, the bioceramics might be used as an inert scaffold. The history related to calcium orthophosphates dates back to 1770 [3] the mistaken for other minerals, such as beryl, tourmaline, chrysolite, amethyst, fluorite, etc., [1, 4, 5]. Currently, apatite is the term for a group of minerals with the same crystallographic structure and older history till 1950 could be delivered somewhere else in the published literature [6, 7]. On the basis of thorough literature survey of HAp, since 1950 in connection to its properties, production, composition and its applications were extensively studied and its usage in medicinal disciplines contributes many breakthroughs in contemporary technological developments in consideration with the interaction of materials on active species [8]. In the origin, HAp was used for grafting, which might not have reaction with neighbouring living cells. Far ahead, the development would change to the responsive nature of the material, also for the growth of bone the reactive material pretends as a conductive scaffold [7]. In recent trend, developing fabrication technology with the dawn of recognizing of regenerative medicinal growth in the field of nanotechnology and have transformed the appearance of bioceramics to a dissimilar facet [9–14].

S. No.	Methods/techniques	Outcome	Drawback	Refs.
1.	Dry	Well-crystallized	High temperature (1050°C in air)	[15, 16]
2.	Wet	High-yield, cost-effective, simple technique, and suitable for various pressure conditions	Non-crystalline and impure phase	[17]
3.	Co-precipitation	Crystalline, high-yield, cost-effective, template-assisted & various temperatures conditions	Requires high temperature annealing to yield product	[18–20]
4.	Sol–gel	Simple technique, low cost, crystalline nature	Dependent on solvent, the temperature and pH	[21, 22]
5.	Emulsion	More efficient, simple and particle agglomeration is less, Suitable for various surfactants, temperature conditions.	Dependent on ratio of aqueous and organic phases, pH and temperature	[23, 24]
6.	Hydrolysis	Simple technique, particle agglomeration is little high, sources are texture dependent	Precursors depend strongly on pH and temperature	[25]
7.	Hydrothermal	Highly crystalline micro or nano-sized structures, well-controlled morphology and porosity	Requires constant and uninterrupted temperature and pressure conditions	[26, 27]
8.	Alternate energy input (low-energy plasma spray)	Uniform thickness, good crystallinity, well-controlled morphology, porosity, micro hardness, and fracture toughness	Requires constant, uninterrupted temperature and pressure conditions. High temp. withstanding substrates and good cleaning process	[28]
9.	Microwave (MW)-assisted	Yield of perfectly, highly crystalline, homogeneous size, porosity and morphology	Requires constant, uninterrupted temperature conditions to yield product	[29]
10.	Ball-milling	Simplicity, reproducibility, and large-scale production	Requires high temperature annealing to yield product and little agglomeration	[30]
11.	Sonochemical	Nanosized products, elicits perfect control of morphology, porosity and size	Requires constant, uninterrupted temperature and pressure conditions.	[31]
12.	Others: a. Solvothermal process b. Spray pyrolysis	Yield of perfectly homogeneous size crystalline morphology	Requires organic solvents and hot zone of an electric furnace	[32, 33]

Table 1. Shows the key methods for the synthesis of hydroxyapatite (HAp).

Hydroxyapatite, HAp is an elementary calcium phosphate, and its chemical formula is $Ca_{10}(PO_4)_6(OH)_2$ present as main material of teeth, bones and mineral certainly with high bi-affinity. It is composited by below methods, and it is also applicable in various fields including biomaterials. In the meantime, amorphous HAp is no longer stable and could dissolve reliant on usage environment; a sintered body has been effectively used as a material in general. And, the sintered body could not dissolve so much owing to its high crystallinity. Because fusion and grain growth of each particle arose on its process stage are foreseeable, it has been hard to control configuration and grain diameter on a nanoscale impartial like initial particles of amorphous HAp. The synthesis of HAp, with its numerous morphologies, structures, and textures, has enthusing a prodigious deal of interest in academic and industrial research for numerous heterogeneous catalysis applications. In the past three decades, a numerous synthetic routes for producing HAp powders have been developed. Productions of HAp powders are classified under four different methods are enumerated in **Table 1** [34].

2. Topical advancements in reformative medicinal treatments in the new prospects of application of nanotechnology

HAp is considered as bioceramics that signifies the enormous amount of regenerative scion material persisting in the flea market. HAp is analogous to the bony-like apatite structure and is considered to be an important inorganic constituent for bone. However, in the organic matrix HAp is circumscribed, so that the existences of HAp in the normal bone in the form of extra inorganic trace elements [3]. Ailments related to the ablative and bone surgical treatment known as the abscission or removing a part of the bone, which ultimately needs renovation through various available measures. Since, the HAp has found increasing demand in regenerative medicine as a possible auxiliary material second to auto graft. HAp could also be used in occurrences, wherever the defects or voids present in bone. This process leads to curing of blocks, or beads by employing powders of the mineral being positioned into or on the defected parts of bone. From the time when it is bioactive, it reassures the bone to spot on the problem for further orientation of growth and this procedure may perhaps be an alternate to bone or dental implants, means that it can integrate into bone or dental structures and support growth with the no breaking down or dissolving in the human body. Though, HAp is still used for this purpose today and it is also applicable for other purposes too. Numerous advancements in nanotechnology oriented reformative medicine for the overhaul or improvement of dented tissues function in several organ systems. However, most studies concern the goings-on of topical advancements in nanomaterials used in regenerative medicinal treatments [35], as summarized in **Table 2**, with some more literatures in HAp, on the basis of regenerative medicine in various organ systems.

Applications of nanotechnology in regenerative medicine would require the entire prospective to reform tissue repair and regeneration [35]. Till now, to trigger the regeneration process the growth of impeccable nanomaterials accomplished of transfer signals to the diseased or damaged cells and tissues it remnants a dare. By employing nano-HAp based materials in regenerative medicine is a material of significant relate to the safety in relations to human health aspects, for the reason that this area is still in its developing platform. Erstwhile to human health based applications, a systematic research work in relevance to the noxious effect of these nanomaterials would be carrying through in excessive manner. In conclusion, at the nanoscale level to make acquainted about the original mechanisms of cell-biomaterial surface interfaces, and further implement the findings from bench to bedside, a manageable teamwork flanked by the scientists and clinicians is of highly necessary for the societal benign.

S. No.	Body part	Nanomaterials	Outcome (type of study)	Refs.
1.	Bone	Poly(epsilon caprolactone)	Improved cell attachment, proliferation, differentiation, and mineralization of osteoblasts (in vitro)	[36]
			Lineage restriction of progenitor cells by topographical cues (in vitro)	[37]
		Nanoscaled calcium phosphate	Large-sized blood vessel infiltration leads to bone formation (in vivo; canines)	[38]
		HAp-coated titanium	Enhanced and accelerated osseoimplant integration (in vivo; rats)	[39]
		Hybrid biomimetic collagen-hydroxyapatite composites	Crosslinking reactions for hard tissue engineering application with designed bioactive properties	[40]
		Nanostructured beta tri-calcium phosphate-coated over poly (lactic acid)	Enhanced osteoconductivity of scaffold (in vitro) and heterotrophic bone formation (in vivo; rabbits)	[41]
		Carbon nanotubes	Extracellular matrix calcification (in vitro); lamellar bone regeneration (in vivo; mice)	[42]
			Porous bone formation in bone defect (in vivo; rats)	[43]
		Silica nanofibers	Proliferation and maturation of MG63 cells (in vitro)	[44]
2.	Cartilage	Pentosan poly sulfate in poly (ethylene glycol) HA	Formation of cartilage like tissues by mesenchymal progenitor cells (in vitro)	[45]
		PVA/PCL [poly(vinyl alcohol) poly(caprolactone)]	Proliferation and chondrogenic differentiation of MSCs (in vitro); improved healing of cartilage defects (in vivo; rabbits)	[46]
		3D porous polycaprolactone (PCL)-hydroxyapatite (HAp) scaffold combined with MC	Improves the biological performance of 3D PCL-HAp scaffold	[47]
		POSS–PCU [polyhedral oligomeric silsesquioxane with polycarbonate polyurethane]	Enhanced survival, proliferation, and chondrogenic differentiation of adipose tissue derived stem cells (in vitro)	[48]
			Enhanced growth and proliferation of nasoseptal chondrocytes (in vitro)	[49]
3.	Peripheral nervous system	Electrospun collagen/poly (lactic-co-glycolic acid)	Axon regeneration, myelination, and action potential propagation (in vivo; rats)	[50]
		Poly(L-lactide-co-glycolide)/chitosan/hydroxyapatite(PLGA/chitosan/HAp)	In vivo application of PLGA/chitosan/HAp conduits for nerve regeneration	[51]
		POSS–PCU–MWCNT	Novel biomaterial capable of electronic interfacing with tissue holds potential to promote nerve regeneration	[52]

S. No.	Body part	Nanomaterials	Outcome (type of study)	Refs.
4.	Central nervous system	Small interfering ribonucleic acid (Si-RNA) chitosan nanoparticles	Increased delivery of drugs by crossing BBB (blood–brain barrier) (in vivo; rats)	[53]
		Nano-HAPs on the growth of human glioma U251 and SHG44 cells in vitro and in vivo	Nano-HAPs have an obvious antineoplastic function in vitro and in vivo. It reduces the poisonous, adverse reactions to 1,3-bis(2-chloroethyl)-1-nitrosourea (BCNU), strongly cooperate with certain other chemotherapy drugs, decrease the toxicity, and might become a new clinical antineoplastic drug.	[54]
5.	Myocardial tissue/myocardial infarction (MI)	Insulin-like growth factor-1 (IGF-1) with poly(lactic-co-glycolic acid)	Increased protein kinase B phosphorylation and reduced infarct size (in vivo; mice)	[55]
		Electrospun (hb/gel/fb) [poly(hemoglobin/gelatin/fibrinogen)]	Cardiomyogenic differentiation of mesenchymal stem cells (MSCs) (in vitro)	[56]
		PGS [poly(glycerol sebacate)]	Increased transplant cell retention and survival (in vitro)	[57]
		Gold nanoparticles-loaded hybrid nanofibers	Cardiomyogenic differentiation of MSCs; superior biological and functional properties (in vitro)	[58]
		Calcium hydroxyapatite–based dermal filler into the infarct	Injection of an acellular dermal filler into an MI immediately after coronary occlusion reduces early infarct expansion and limits chronic LV remodeling.	[59]
6.	Skin	Silver nanoparticles	Reduced inflammation and promotion of wound healing (in vitro)	[60]
		Plasma-treated electrospun poly(lactic-acid) co-poly(epsilon caprolactone), and gelatin	Increased fibroblast proliferation and collagen secretion (in vitro)	[61]
		Titanium abutment (control) and one HA-coated abutment (case) interface	The HAp-coated abutment can achieve integration with the surrounding skin.	[62]
		Rosette nanotubes with PHeMA [poly(2-hydroxyethyl methacrylate]	Increased keratinocyte and fibroblast proliferation (in vitro)	[63]
7.	Eye	Polydimethylsiloxane	Topographical cue for formation of functioning corneal endothelium (in vitro)	[64]
		HAp, polytetrafluoroethylene (PTFE), polyhydroxyethyl methacrylate (HEMA), and glass (control)	Improving the initial cell adhesion environment in the skirt element of keratoprostheses may enhance tissue integration and reduce device failure rates.	[65]
		Super paramagnetic nanoparticles	Increased gene expression and neurite growth, subcellular organelle localization, and nano therapeutics delivery (in vitro)	[66]

S. No.	Body part	Nanomaterials	Outcome (type of study)	Refs.
8.	Lung	Deferoxamine	Regeneration of microvascular anastomosis in airways (in vivo; mice)	[67]
		HAPNs in both A549 and 16HBE cells	HAPNs might be a promising agent or mitochondria-targeted delivery system for effective lung cancer therapy.	[68]
		101F6 (tumor suppressor gene) nanoparticles	Increased tumor cell lysis (in vitro and in vivo; mice)	[69]

Courtesy: Reproduced from Ref. [35] with permission from Dove Medical Press, copyright 2014.

Table 2. Topical advancements in nanomaterials used in regenerative medicinal treatments [35].

3. Conclusion

In summary, hydroxyapatite is one of the well-studied biomaterials in the medical field for its established biocompatibility and for being the main content of the mineral part of bone, teeth and various organ systems. However the fact demonstrates that it has been more imperious towards ground-breaking research against novel medical applications for the cause of the society. It has all the typical topographies of biomaterials, such as, bioactive, biocompatible, non-toxic, osteoconductive, non-immunogenic, non-inflammatory, bioceramic coatings, bone void fillers for orthopedics, dental implant coating, HAp thin films, and resemblance to the inorganic component of human beings. In the midst of the major remarkable progress are in various fields of molecular biology, biochemistry, bioinformatics, microbiology, genetics, cytometry, medical diagnostics, drug & gene delivery, and the addition of nanotechnology are the most important worldwide challenges so far. The dispute of novel spectroscopic/microscopical innovation contains interdisciplinary areas that might endure to be enhanced for these innovative global developments in x-ray imaging, spectral imaging, time-correlated single-photon counting, fluorescence quenching, endo- and exo-thermic reaction rates, kinetic chemical reaction rates, *In vitro* and *In vivo* studies, visual implants, neurology and non-invasive optical biopsy. Thus, studies towards unique nano-hydroxyapatite used in regenerative medicinal treatments might give way to mechanisms of cell-biomaterial relations at the nanoscale level that may feasibly turn out to be the upcoming forerunners to human applications in the embryonic stage.

Acknowledgements

All authors contributed towards data analysis, drafting and revising the paper and agree to be accountable for all aspects of the work. The authors apologize for inadvertent omission of any pertinent references.

Conflict of interest

The authors declare that there is no conflict of interests regarding the publication of this paper.

Author details

Jagannathan Thirumalai

Address all correspondence to: thirumalaijg@gmail.com

Department of Physics, School of Electrical & Electronics Engineering, SASTRA Deemed University, Srinivasa Ramanujan Centre, Kumbakonam, Tamil Nadu, India

References

[1] Dorozhkin SV. A detailed history of calcium orthophosphates from 1770s till 1950. Materials Science and Engineering: C. 2013;**33**(1):3085-3110. DOI: 10.1016/j.msec.2013.04.002

[2] Dubok VA. Bioceramics – Yesterday, today, tomorrow. Powder Metallurgy and Metal Ceramics. 2000;**39**(7-8):381-394. DOI: 10.1023/A:1026617607548

[3] Hench LL, Tompson I. Twenty-first century challenges for biomaterials. Journal of Royal Society Interface. 2010;**7**(Suppl 4):S379-S391. DOI: 10.1098/rsif.2010.0151.focus

[4] Dorozhkin SV. Calcium orthophosphates in nature, Biology and Medicine. Materials. 2009;**2**(2):399-498. DOI: 10.3390/ma2020399

[5] Dorozhkin SV. Calcium orthophosphates: Occurrence, properties, biomineralization, pathological calcification and biomimetic applications. Biomatter. 2011;**1**(2):121-164. DOI: 10.4161/biom.18790

[6] Weiner S, Wagner HD. The Material Bone: Structure-mechanical function relations. Annual Review of Materials Science. 1998;**28**(1):271-298. DOI: 10.1146/annurev.matsci.28.1.271

[7] Dorozhkin SV. Calcium Orthophosphates Applications in Nature, Biology and Medicine. 1st ed. Boca Raton: Pan Stanford Publishing; 2012. p. 870. DOI: 10.4032/9789814364171

[8] LeGeros RZ. Calcium phosphate-based osteoinductive materials. Chemical Reviews. 2008;**108**(11):4742-4753. DOI: 10.1021/cr800427g

[9] Ohgushi H, Dohi Y, Tamai S, Tabata S. Osteogenic differentiation of marrowstromal stem cells in porous hydroxyapatite ceramics. Journal of Biomedical Materials Research. 1993;**27**(11):1401-1407. DOI: 10.1002/jbm.820271107

[10] Ripamonti U, Crooks J, Khoali L, Roden L. Te induction of bone formation by coral-derived calcium carbonate/hydroxyapatite constructs. Biomaterials. 2009;**30**(7):1428-1439. DOI: 10.1016/j.biomaterials.2008.10.065

[11] Ripamonti U, Richter PW, Nilen RWN, Renton L. Te induction of bone formation by smart biphasic hydroxyapatite tricalcium phosphate biomimetic matrices in the non-human primate Papio ursinus. Journal of Cellular and Molecular Medicine. 2008;**12**(6b):1-15. DOI: 10.1111/j.1582-4934.2008.00312.x

[12] Yuan H, Kurashina K, de Bruijn JD, Li Y, de Groot K, Zhang X A preliminary study on osteoinduction of two kinds of calcium phosphate ceramics. Biomaterials 1999;**20**(19): 1799-1806. DOI: 10.1016/S0142-9612(99)00075-7

[13] Habibovic P, Gbureck U, Doillon CJ, Bassett DC, van Blitterswijk CA, Barralet JE. Osteoconduction and osteoinduction of low-temperature 3D printed bioceramic implants. Biomaterials. 2008;**29**(7):944-953. DOI: 10.1016/j.biomaterials.2007.10.023

[14] Ripamonti U, Roden LC, Renton LF. Osteoinductive hydroxyapatite-coated titanium implants. Biomaterials. 2012;**33**(15):3813-3823. DOI: 10.1016/j.biomaterials.2012.01.050

[15] Korber F, Trömel GZ. The formation of HAP through a solid state reaction between tri and tetra – calcium phosphates. Journal of The Electrochemical Society. 1932;**38**:578-580

[16] Guo X, Yan H, Zhao S, Li Z, Li Y, Liang X. Effect of calcining temperature on particle size of hydroxyapatite synthesized by solid-state reaction at room temperature. Advanced Powder Technology. 2013;**24**(6):1034-1038. DOI: 10.1016/j.apt.2013.03.002

[17] Pramanik S, Agarwal AK, Rai KN, Garg A. Development of high strength hydroxyapatite by solid-state-sintering process. Ceramics International. 2007;**33**(3):419-426. DOI: 10.1016/j.ceramint.2005.10.025

[18] Ikoma T, Yamazaki A, Nakamura S, Akao M. Preparation and structure refinement of monoclinic hydroxyapatite. Journal of Solid State Chemistry. 1999;**144**(2):272-276. DOI: 10.1006/jssc.1998.8120

[19] Tao J, Jiang W, Pan H, Xu X, Tang R. Preparation of large-sized hydroxyapatite single crystals using homogeneous releasing controls. Journal of Crystal Growth. 2007;**308**(1):151-158. DOI: 10.1016/j.jcrysgro.2007.08.009

[20] Swain SK, Sarkar D. A comparative study: Hydroxyapatite spherical nanopowders and elongated nanorods. Ceramics International. 2011;**37**(2):2927-2930. DOI: 10.1016/j.ceramint.2011.03.077

[21] Chen J, Wang Y, Chen X, Ren L, Lai C, He W, Zhang Q. A simple sol-gel technique for synthesis of nanostructured hydroxyapatite, tricalcium phosphate and biphasic powders. Materials Letters. 2011;**65**(12):1923-1926. DOI: 10.1016/j.matlet.2011.03.076

[22] Rajabi-Zamani AH, Behnamghader A, Kazemzadeh A. Synthesis of nanocrystalline carbonated hydroxyapatite powder via nonalkoxide sol–gel method. Materials Science and Engineering: C. 2008;**28**(8):1326-1329. DOI: 10.1016/j.msec.2008.02.001

[23] Shum HC, Bandyopadhyay A, Bose S, Weitz DA. Double emulsion droplets as microreactors for synthesis of mesoporous hydroxyapatite. Chemistry of Materials. 2009;**21**(22): 5548-5555. DOI: 10.1021/cm9028935

[24] Zhou W, Wang M, Cheung W, Guo B, Jia D. Synthesis of carbonated hydroxyapatite nanospheres through nanoemulsion. Journal of Materials Science. Materials in Medicine. 2008;**19**(1):103-110. DOI: 10.1007/s10856-007-3156-9

[25] Sturgeon JL, Brown PW. Effects of carbonate on hydroxyapatite formed from CaHPO4 and Ca4(PO4)2O. Journal of Materials Science. Materials in Medicine. 2009;**20**(9):1787-1794. DOI: 10.1007/s10856-009-3752-y

[26] Lin K, Liu X, Chang J, Zhu Y. Facile synthesis of hydroxyapatite nanoparticles, nanowires and hollow nano-structured microspheres using similar structured hard-precursors. Nanoscale. 2011;**3**(8):3052-3055. DOI: 10.1039/c1nr10334b

[27] Thirumalai J, Chandramohan R, Vijayan TA. Synthesis, characterization and growth mechanism of dumbbell-shaped Fluoroapatite (FHAp) superstructures. Advanced Science Letters. 2012;**5**(1):118-123. DOI: 10.1166/asl.2012.1925

[28] Garcia-Alonso D, Parco M, Stokes J, Looney L. Low-Energy Plasma Spray (LEPS) deposition of hydroxyapatite/poly-e-Caprolactone biocomposite coatings. Journal of Thermal Spray Technology. 2012;**21**(1):132-143. DOI: 10.1007/s11666-011-9695-0

[29] Farzadi A, Solati-Hashjin M, Bakhshi F, Aminian A. Synthesis and characterization of hydroxyapatite/β-tricalcium phosphate nanocomposites using microwave irradiation. Ceramics International. 2011;**37**(1):65-71. DOI: doi.org/10.1016/j.ceramint.2010.08.021

[30] El Briak-Benabdeslam H, Ginebra MP, Vert M, Boudeville P. Wet or dry mechanochemical synthesis of calcium phosphates? Influence of the water content on DCPD–CaO reaction kinetics. Acta Biomaterialia. 2008;**4**(2):378-386. DOI: 10.1016/j.actbio.2007.07.003

[31] Giardina MA, Fanovich MA. Synthesis of nanocrystalline hydroxyapatite from $Ca(OH)_2$ and H_3PO_4 assisted by ultrasonic irradiation. Ceramics International. 2010;**36**(6):1961-1969. DOI: 10.1016/j.ceramint.2010.05.008

[32] Smolen D, Chudoba T, Malka I, Kedzierska A, Lojkowski W, Swieszkowski W, Kurzydlowski KJ, Kolodziejczyk-Mierzynska M, Lewandowska-Szumie M. Highly biocompatible, nanocrystalline hydroxyapatite synthesized in a solvothermal process driven by high energy density microwave radiation. International Journal of Nanomedicine. 2013;**8**(1):653-668. DOI: 10.2147/IJN.S39299

[33] Cho JS, Lee JC, Rhee SH. Effect of precursor concentration and spray pyrolysis temperature upon hydroxyapatite particle size and density. Journal of Biomedical Materials Research. Part B, Applied Biomaterials. 2016;**104**(2):422-430. DOI: 10.1002/jbm.b.33406

[34] Fihri A, Len C, Varma RS, Solhy A. Hydroxyapatite: A review of syntheses, structure and applications in heterogeneous catalysis. Coordination Chemistry Reviews. 2017;**347**(9):48-76. DOI: 10.1016/j.ccr.2017.06.009

[35] Chaudhury K, Kumar V, Kandasamy J, RoyChoudhury S. Regenerative nanomedicine: Current perspectives and future directions. International Journal of Nanomedicine. 2014;**9**(1):4153-4167. DOI: 10.2147/IJN.S45332

[36] Wang J, Valmikinathan CM, Liu W, Laurencin CT, Yu X. Spiral structured, nanofibrous, 3D scaffolds for bone tissue engineering. Journal of Biomedical Materials Research. Part A. 2010;**93**(2):753-762. DOI: 10.1002/jbm.a.32591

[37] Cassidy JW, Roberts JN, Smith CA, Robertson M, White K, Biggs MJ, Oreffo RO, Dalby MJ. Osteogenic lineage restriction by osteoprogenitors cultured on nanometric grooved surfaces: The role of focal adhesion maturation. Acta Biomaterialia. 2014;**10**(2):651-660. DOI: 10.1016/j.actbio.2013.11.008

[38] Appleford MR, Oh S, Oh N, Ong JL. In vivo study on hydroxyapatite scaffolds with trabecular architecture for bone repair. Journal of Biomedical Materials Research. Part A. 2009;**89**(4):1019-1027. DOI: 10.1002/jbm.a.32049

[39] Yamada M, Ueno T, Tsukimura N, et al. Bone integration capability of nanopolymorphic crystalline hydroxyapatite coated on titanium implants. International Journal of Nanomedicine. 2012;**7**(1):859-873. DOI: 10.2147/IJN.S28082

[40] Krishnakumar GS, Gostynska N, Dapporto M, Campodoni E, Montesi M, Panseri S, Tampieri A, Kon E, Marcacci M, Sprio S, Sandric M. Evaluation of different crosslinking agents on hybrid biomimetic collagen-hydroxyapatite composites for regenerative medicine. International Journal of Biological Macromolecules. 2017;**106**:739-748. DOI: 10.1016/j.ijbiomac.2017.08.076

[41] Cao L, Duan PG, Wang HR, et al. Degradation and osteogenic potential of a novel poly(lactic acid)/nano-sized β-tricalcium phosphate scaffold. International Journal of Nanomedicine. 2012;**7**(1):5881-5888. DOI: 10.2147/IJN.S38127

[42] Lobo AO, Siqueira IA, Das Neves MF, Marciano FR, Corat EJ, Corat MA. In vitro and in vivo studies of a novel nanohydroxyapatite/superhydrophilic vertically aligned carbon nanotube nanocomposites. Journal of Materials Science. Materials in Medicine. 2013;**24**(7):1723-1732. DOI: 10.1007/s10856-013-4929-y

[43] Hirata E, Ménard-Moyon C, Venturelli E, et al. Carbon nanotubes functionalized with fibroblast growth factor accelerate proliferation of bone marrow-derived stromal cells and bone formation. Nanotechnology. 2013;**24**(43):435101. DOI: 10.1088/0957-4484/24/43/435101

[44] Ravichandran R, Gandhi S, Sundaramurthi D, Sethuraman S, Krishnan UM. Hierarchical mesoporous silica nanofibers as multifunctional scaffolds for bone tissue regeneration. Journal of Biomaterials Science. Polymer Edition. 2013;**24**(17):1988-2005. DOI: 10.1080/09205063.2013.816930

[45] Frith JE, Cameron AR, Menzies DJ, et al. An injectable hydrogel incorporating mesenchymal precursor cells and pentosan polysulphate for intervertebral disc regeneration. Biomaterials. 2013;**34**(37):9430-9440. DOI: 10.1016/j.biomaterials.2013.08.072

[46] Shafee A, Soleimani M, Chamheidari GA, et al. Electrospun nanofberbased regeneration of cartilage enhanced by mesenchymal stem cells. Journal of Biomedical Materials Research. Part A. 2011;**99**(3):467-478. DOI: 10.1002/jbm.a.33206

[47] Yao Q, Wei B, Liu N, Li C, Guo Y, Shamie AN, Chen J, Tang C, Jin C, Xu Y, Bian X, Zhang X, Wang L. Chondrogenic regeneration using bone marrow clots and a porous polycaprolactone-hydroxyapatite scaffold by three-dimensional printing. Tissue Engineering. Part A. 2015;**21**(7-8):1388-1397. DOI: 10.1089/ten.TEA.2014.0280

[48] Guasti L, Vagaska B, Bulstrode NW, Seifalian AM, Ferretti P. Chondrogenic differentiation of adipose tissue-derived stem cells within nanocaged POSS-PCU scaffolds: A new tool for nanomedicine. Nanomedicine. 2014;**10**(2):279-289. DOI: 10.1016/j.nano.2013.08.006

[49] Oseni AO, Butler PE, Seifalian AM. The application of POSS nanostructures in cartilage tissue engineering: The chondrocyte response to nanoscale geometry. Journal of Tissue Engineering and Regenerative Medicine. 2015;**9**(11):E27-E38. DOI: 10.1002/term.1693

[50] Ouyang Y, Huang C, Zhu Y, Fan C, Ke Q. Fabrication of seamless electrospun collagen/PLGA conduits whose walls comprise highly longitudinal aligned nanofibers for nerve regeneration. Journal of Biomedical Nanotechnology. 2013;**9**(6):931-943. DOI: 10.1166/jbn.2013.1605

[51] Shen X, Ruan J, Zhou Z, Zeng Z, Xie L. Evaluation of PLGA/chitosan/HA conduits for nerve tissue reconstruction. Journal of Wuhan University of Technology-Materials Science Edition. 2009;**24**(4):566-570. DOI: 10.1007/s11595-009-4566-y

[52] Antoniadou EV, Ahmad RK, Jackman RB, Seifalian AM. Next generation brain implant coatings and nerve regeneration via novel conductive nanocomposite development. In: Nigel Lovell, editor. Conference proceedings IEEE Engineering in Medicine and Biology Society, EMBC, 2011 annual international conference of the IEEE; 30 August 3 – September 2011; Boston, MA: IEEE; 2011. pp. 3253-3257. DOI: 10.1109/IEMBS.2011.6090884

[53] Malmo J, Sandvig A, Vårum KM, Strand SP. Nanoparticle mediated P-glycoprotein silencing for improved drug delivery across the blood–brain barrier: A siRNA-chitosan approach. PLoS One. 2013;**8**(1):e54182. DOI: 10.1371/journal.pone.0054182

[54] Chu SH, Feng DF, Ma YB, Li Z-Q. Hydroxyapatite nanoparticles inhibit the growth of human glioma cells in vitro and in vivo. International Journal of Nanomedicine. 2012;7(7):3659-3666. DOI: 10.2147/IJN.S33584

[55] Chang MY, Yang YJ, Chang CH, et al. Functionalized nanoparticles provide early cardioprotection after acute myocardial infarction. Journal of Controlled Release. 2013;**170**(2):287-294. DOI: 10.1016/j.jconrel.2013.04.022

[56] Ravichandran R, Seitz V, Reddy Venugopal J, et al. Mimicking native extracellular matrix with phytic acid-crosslinked protein nanofibers for cardiac tissue engineering. Macromolecular Bioscience. 2013;**13**(3):366-375. DOI: 10.1002/mabi.201200391

[57] Ravichandran R, Venugopal JR, Sundarrajan S, Mukherjee S, Sridhar R, Ramakrishna S. Minimally invasive injectable short nanofibers of poly(glycerol sebacate) for cardiac tissue engineering. Nanotechnology. 2012;**23**(38):385102. DOI: 10.1088/0957-4484/23/38/385102

[58] Ravichandran R, Sridhar R, Venugopal JR, Sundarrajan S, Mukherjee S, Ramakrishna S. Gold nanoparticle loaded hybrid nanofibers for cardiogenic differentiation of stem cells for infarcted myocardium regeneration. Macromolecular Bioscience. 2014;**14**(4):515-525. DOI: 10.1002/mabi.201300407

[59] Ryan LP, Matsuzaki K, Noma M, Jackson BM, Eperjesi TJ, Plappert TJ, St. John-Sutton MG, Gorman JH, Gorman RC. Dermal filler injection: A novel approach for limiting infarct expansion. The Annals of Thoracic Surgery. 2009;**87**(1):148-155. DOI: 10.1016/j.athoracsur.2008.09.028

[60] Wu J, Zheng Y, Song W, et al. Situ synthesis of silver-nanoparticles/bacterial cellulose composites for slow-released antimicrobial wound dressing. Carbohydrate Polymers. 2014;**102**(2):762-771. DOI: 10.1016/j.carbpol.2013.10.093

[61] Chandrasekaran AR, Venugopal J, Sundarrajan S, Ramakrishna S. Fabrication of a nanofibrous scaffold with improved bioactivity for culture of human dermal fibroblasts for skin regeneration. Biomedical Materials. 2011;**6**(1):015001

[62] Hoof MV, Wigren S, Duimel H, Savelkoul PHM, Flynn M, Stokroos RJ. Can the hydroxyapatite-coated skin-penetrating abutment for bone conduction hearing implants integrate with the surrounding skin? Frontiers in Surgery. 2015;**2**(1):1-8. DOI: 10.3389/fsurg.2015.00045

[63] Sun L, Li D, Hemraz UD, Fenniri H, Webster TJ. Self-assembled rosette nanotubes and poly (2-hydroxyethyl methacrylate) hydrogels promote skin cell functions. Journal of Biomedical Materials Research. Part A. 2014;**102**(10):3446-3451. DOI: 10.1002/jbm.a.35008

[64] Teo BK, Goh KJ, Ng ZJ, Koo S, Yim EK. Functional reconstruction of corneal endothelium using nanotopography for tissue-engineering applications. Acta Biomaterialia. 2012;**8**(8):2941-2952. DOI: 10.1016/j.actbio.2012.04.020

[65] Mehta JS, Futter CE, Sandeman SR, Faragher RGAF, Hing KA, Tanner KE, Allan BDS. Hydroxyapatite promotes superior keratocyte adhesion and proliferation in comparison with current keratoprosthesis skirt materials. Journal of Ophthalmology. 2005;**89**(10):1356-1362. DOI: 10.1136/bjo.2004.064147

[66] Steketee MB, Moysidis SN, Jin XL, Weinstein JE, Pita-Thomas W, Hemalatha BR, Siraj I, Jeffrey LG. Nanoparticle-mediated signalling endosome localization regulates growth cone motility and neurite growth. Proceedings of the National Academy of Sciences of the United States of America. 2011;**108**(47):19042-19047. DOI: 10.1073/pnas.1019624108

[67] Jiang X, Malkovskiy AV, Tian W, Sung YK, Sun W, Hsu JL, Manickam S, Wagh D, Joubert LM, Semenza GL, Rajadas J, Nicolls MR. Promotion of airway anastomotic microvascular regeneration and alleviation of airway ischemia by deferoxamine nanoparticles. Biomaterials. 2014;**35**(2):803-813. DOI: 10.1016/j.biomaterials.2013.09.092

[68] Sun Y, Chen Y, Ma X, Yuan Y, Liu CS, Kohn J, Qian JC. Mitochondria-targeted hydroxyapatite nanoparticles for selective growth inhibition of lung cancer in vitro and in vivo. ACS Applied Materials & Interfaces. 2016;**8**(39):25680-25690. DOI: 10.1021/acsami.6b06094

[69] Ohtani S, Iwamaru A, Deng W, et al. Tumor suppressor 101F6 and ascorbate synergistically and selectively inhibit non-small cell lung cancer growth by caspase-independent apoptosis and autophagy. Cancer Research. 2007;**67**(13):6293-6303. DOI: 10.1158/0008-5472

Chapter 2

Hydroxyapatite Production by an Intensification Process

Benedetta de Caprariis, Angelo Chianese,
Marco Stoller and Nicola Verdone

Additional information is available at the end of the chapter

http://dx.doi.org/10.5772/intechopen.71775

Abstract

Hydroxyapatite (HAP) is a worthwhile compound for its biomedical applications. Nanoparticles (NPs) and nanostructured HAP scaffolds promote and intensify the interaction between artificial material and natural bone due to their high surface/volume ratio. In this chapter, first, the technique for the production of HPA nanoparticles smaller than 100 nm is addressed. It consists of the use of a rotating disk reactor to optimize the reaction-precipitation process. The centrifugal force dispersed into the liquid layer over the disk surface enables the attainment of micromixing conditions between the reagents and maximizes the reaction rate as a consequence. The reaction between calcium chloride and ammonium phosphate in the presence of ammonium hydroxide was adopted. NPs minimum size, equal to 78 μm, was obtained using a rotational velocity of 147 rad/s and feeding points of reagents 3 cm from the disk center. A computational fluid dynamics (CFD) model of the liquid layer was specifically developed for the interpretation of the obtained experimental results on the production of pure HAP. In the second part of the chapter, the feasibility of producing Mg^{2+} doped hydroxyapatite (Mg-HAP) by adding $MgCl_2$ and using the same technique is reported. Satisfactory results were obtained: nanoparticles were between 50 and 70 μm in size and Mg^{2+}/Ca^{2+} molar ratio was equal to 0.06, according to the composition target.

Keywords: hydroxyapatite, precipitation, spinning disk, micromixing, Mg-doped hydroxyapatite

1. Introduction

In recent years, several chemistry-based processing routes have been reported for preparing hydroxyapatite (HAP) powders. Nanoparticles with several morphologies have been synthesized by means of solid-state reaction, emulsion techniques, sol-gel and hydrothermal method [1];

however, the wet chemical precipitation method was proven to be one of the easiest ways for preparing HAP powders.

Wet chemical precipitation consists of a chemical reaction followed by the precipitation of the reaction product, which is a very sparingly solute. The overall process can be distinguished in several stages: the mixing among the reagents in liquid phase or gas-liquid phase, the reaction, the nucleation of solid particles as soon as the supersaturation has overcome the metastable limit, the growth and the aggregation of the solid particles. In order to produce solid nanoparticles, it is necessary to maximize nucleation rate and to minimize both growth and aggregation rate. In this respect, the key factor is the intensification of the mixing process. In fact, if the local mixing among the reagent streams, called micromixing, is very effective, the reagents locally attain the maximum concentration. As a consequence, the reaction rate takes place at the maximum rate, the maximum concentration of the required product, that is, its maximum supersaturation, is achieved and the nucleation rate is optimized.

According to the attained supersaturation, homogeneous or heterogeneous nucleation takes place. If the micromixing conditions are intense enough, the micromixing time is smaller than 1 ms and homogeneous nucleation becomes dominant with respect to heterogeneous one. In general, when the needed average crystal size is around one micron or smaller with tight crystal size distribution, homogeneous nucleation is preferred to heterogeneous nucleation. The major objective in this matter is, thus, to develop a precipitation reactor, which intensifies the mixing between the reagents much more than in a stirred reactor, in order to produce nanoparticles in a very narrow size range. For this purpose, microreactors, so-called T-mixer reactor and rotating disk reactor are adopted. All these reactors may assure conditions of micromixing, which usually induces homogeneous nucleation. More recently, the use of tubular microreactors have been proposed. Such apparatuses are characterized by very fast heat and mass transfer and a very small residence time of reagents fed to microstructured devices where the reactions are performed at very controlled conditions [2]. However, the application of this technique at the industrial scale seems to be hard to be proven. The T-mixer is often adopted as premixing device for precipitation-stirred reactors, but cannot represent a practical solution as a reactor stands alone, in particular for its needed high-dispersed energy. The rotating disk reactor requires lower energy amount with respect to the two above-mentioned reactors, may work in continuous mode and is of relatively simple scale-up. The present chapter is focused on the production process of nanoparticles of HAP, not-doped or Mg-doped, by using a spinning disk reactor (SDR).

The first attempt to use the SDR for a precipitation process was afforded by Cafiero et al. [3] by performing the wet precipitation of $BaSO_4$ over a disk rotating between 200 and 1000 rpm. In a subsequent paper, the same authors [4] calculated that at the highest rotational speed a micromixing time between the reagents smaller than 1 ms occurred and particles of barium sulfate around 0.5 μm in size were obtained. Moreover, a comparison between the T-mixer and the SDR performances, leading to the same micromixing, was made and it was shown that T-mixer is much energy consuming with respect to SDR.

Some other works reported the chance to use SDR to produce nanoparticles: Trippa et al. [5] studied the production of calcium carbonate particles from dissolved CO_2; Raston et al. [6]

showed the feasibility of producing b-carotene nanoparticles; Loh et al. [7] showed the potentiality of the spinning disk technology for large manufacture of chitosan nanoparticles; and finally Dabir et al. [8] present an experimental method for large-scale production of silver chloride nanoparticles using SDR.

In all these works, the importance of hydrodynamics of the liquid over the disk surface was emphasized. The prediction of the hydrodynamics over the SDR has been studied by many authors. The hydrodynamic simulation models were often evaluated with respect to the experimental values reported by Burns et al. [9], concerning the measurements of the liquid film thickness over a rotating disk. In the same paper, there was a comparison between the Pigford model and the Nusselt theory and it was shown that this latter cannot satisfactorily predict the liquid profiles over the disk when the inertial forces are higher than the viscous one, that is, for low values of the Ekman number. More recently, Bhatelia et al. [10] studied the prediction of the liquid layer over the rotating disk by a CFD model and they obtained results in good agreement with those reported by Burns. de Caprariis et al. [11] developed a CFD simulation model to predict the hydrodynamics of the reagent solutions over an SDR used for the production of hydroxyapatite. It was clearly shown both the patterns of the reagent streams fed over the disk and the progressively decrease of the reagents concentrations. The first work dealing with the production of nanoparticles in a two-phase solid liquid was that one of Plasari et al. [12]. Nucleation and growth were considered, whereas aggregation of particles was neglected. Some deviations between experimental and simulated values were obtained. More recently, de Caprariis et al. [13] tried to predict the crystallite of HPA produced over the spinning disk by inserting the particle balance equation in the CFD model previously developed by the same authors. Definitely, nowadays much work has to be done on the prediction of nanoparticles produced by means of an SDR.

In this chapter, the effects of several operating parameters (reagent flow rate, rotational speed and feed point location) on the size of pure HAP and Mg-HAP nanoparticles have been investigated. The nanoparticles size minimization for not-doped HAP, and for Mg-doped HAP at a fixed molar ratio Mg^{2+}/Ca^{2+} equal to 0.06, was pursued. The main aim has been to show the feasibility of producing nanoscale HAP nanoparticles by wet chemical synthesis in an SDR, by operating in continuous mode.

2. Description of the SDR system for the production of pure HAP and mg-HAP nanoparticles

Spinning disk reactor (SDR) appears to be a versatile and efficient equipment for the production of nanoparticles by wet chemical synthesis. As soon as chemical precipitation takes place under intensified micromixing conditions, a two-step nucleation-aggregation process mainly determines the size distribution of the produced nanoparticles. In these conditions, almost all the generated supersaturation is quickly consumed by nucleation, and only a small amount of the residual local supersaturation is available for growth and aggregation. First of all, it is interesting to point out the effect of the mixing intensification on the two main phenomena of nucleation and aggregation. Hounslow and Mumtaz [14] have described the two-step process of aggregation: a shear stress orthokinetic collision between two particles and a subsequent cementation of these at their contact point due to internal diffusion of solute ions. In this process, fluid shear appears on the one hand to be in favor of aggregation by increasing the number and intensity of the particle collisions and on the other hand in disagreement, since it reduces the effect of the collisions due to reduced time for bridge formation and its disruptive action on the agglomerated particles. These considerations were derived from experimental runs on the aggregation of nanoparticles of calcium oxalate in a Poiseuille flow crystallizer. In this case, aggregation was reduced by intensified operating conditions, leading mainly to the disruptive action of the fluid shear stress. As a consequence, it appears that the mixing process intensification, enhancing nucleation and reducing aggregation process, leads to the production of smaller nanoparticles in size.

Some works emphasized the great importance of hydrodynamics on the produced particles size, taking into account the feeding points of the reagents over the disk surface. The paper of Parisi et al. [15] on the production of HAP particles pointed out the importance of the feed stream injection points over the disk to achieve specific particle size distributions and yields. Moreover, Stoller et al. [16] showed that the location of the injection points is very important with respect to the aggregation rate and the scaling formation over the reactor surface. In this chapter, the importance of the feed flow rate on the particle agglomeration was also observed, and in fact the density of the produced nanoparticles over the disk surface increases the aggregation rate.

Summarizing, the performances of a given SDR appear to be a function of the disk rotational speed, feed flow rate(s) and location of the injection points. The experimental device used in this work is schematized in **Figure 1**. It consists of a cylindrical case with an inner disk of 8.5 cm in diameter, made by PVC.

Rotational velocity of the disk could be increased up to 147 rad/s, corresponding to 1500 rpm. The reagent solutions were fed over the disk at a distance of 5 mm from the disk surface through tubes, 1 mm in diameter. The position of the reagent injection points over the disk

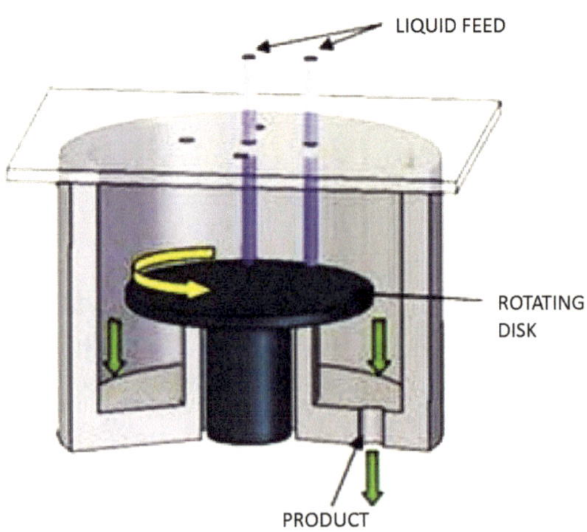

Figure 1. Scheme of the adopted SDR.

surface as a function of the distance from the disk center could be chosen between 0 and 3 cm. The constancy of the feed flow rates of each reagent solution stream was assured by the use of peristaltic pumps.

The produced suspension left the disk in continuous mode from its periphery and then suddenly came out of the cylindrical case. All the experiments were conducted at room temperature.

3. Experimental results

In the following two sections, the major results pertaining to the experimental tests conducted for the synthesis of HAP and Mg-doped HAP nanoparticles are reported. The detailed description of the adopted experimental procedures and the extensive discussions on the obtained results are reported in the cited papers.

3.1. Production process of HAP

In case of pure HAP production, three solutions were fed over the disk surface: the two reagent solutions at a distance from the disk center of 2 or 3 cm and an aqueous solution of ammonium hydroxide at the disk center. In particular, a 10% aqueous solution of NH_4OH at a flow rate of 80 ml/min was fed, whereas the two reagent aqueous streams had both a flow rate of 100 ml/min and a solute mass fraction of 5.6% of $CaCl_2$ and 3.5% of $(NH_4)_2HPO_4$, respectively. The calcium/phosphate (Ca/P) ratio of 1.67, corresponding to stoichiometric HAP, was respected. This condition is considered by many researchers very important in order to obtain nanoparticles of hydroxyapatite with high purity [17].

The reaction takes place between calcium chloride and ammonium phosphate, in the presence of ammonium hydroxide, according to the stoichiometry:

$$10CaCl_2 + 6(NH_4)_2HPO_4 + 8NH_4OH \rightarrow Ca_{10}(PO_4)_6(OH)_2 + 20(NH_4)\,Cl + 6H_2O \quad (1)$$

Ammonium hydroxide is used to attain a pH value equal to 10 and, as a consequence, high yield of the reaction to HAP [17]. Particle size distribution was measured by a dynamic light scattering instrument (DLS, PLUS 90 by Brookhaven) in the range 1–6000 nm. The samples were prepared by dispersing small amounts of the HAP powder in a 25-mL NaOH solution (0.1 M, pH 10) with 0.2 g of the surfactant Twin60 and submitting this suspension to ultrasonication for 15 min.

The experiments were focused on the evaluation of the effects of the rotational speed and of the radial distance of reagent feeding points from the disk center on the size of the HAP particles. In fact, these two parameters strongly affect the local micromixing time and as a consequence the achieved nucleation rate.

The majority of the experimental tests were carried out, feeding the reagents at 2 cm from the disk center and varying the rotational speed between 58 rad/s and 147 rad/s. Furthermore, in order to evaluate the effect of the feeding point, two runs were performed, at constancy of rotational speed of 147 rad/s, by feeding the two reagents at opposite distance of 3 cm from the center.

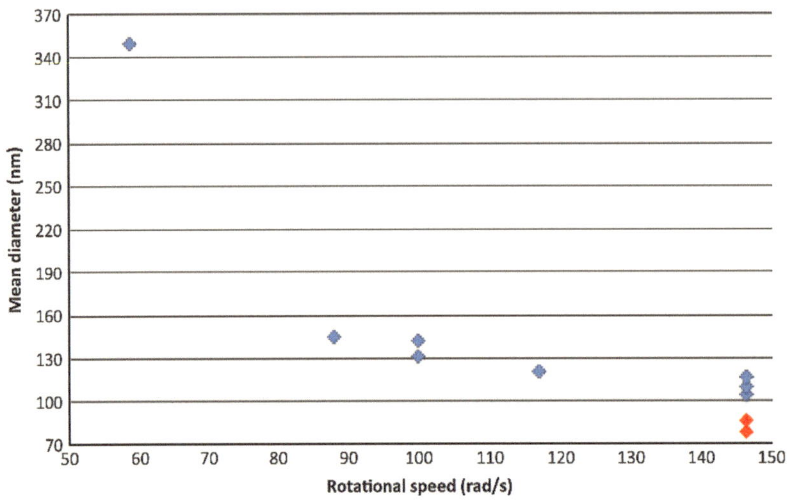

Figure 2. Average size of the produced particles, varying the rotational speed and the injection position. Blue points and red point refer to an injection point at 2 and 3 cm from the disk center, respectively.

The measured particle size as a function of rotational speed and feeding point distance is reported in **Figure 2**. As expected, the mean size of the produced particles is inversely proportional to the local energy dissipation due to the centrifugal force. In fact, this latter becomes higher when the disk rotational velocity increases and the radial position of the feed point approaches the disk edge. The minimum size, equal to 78 nm, was obtained using a rotational velocity of 147 rad/s and feeding points of reagents 3 cm from the disk center.

A typical size distribution measurement of the HAP nanoparticles is reported in **Figure 3**.

It has to be noticed that the produced nanoparticles, even in the nanometers range, are aggregations of single particles around 5 nm in size. This is, in fact, the dimension of a single crystallite estimated using the Scherrer's formula from X-ray diffractometer measurements for particles produced at the maximum rotational speed [13, 18].

The key of a rapid mixing is to produce a region of high turbulent energy dissipation, as a matter of fact that the increase of the rotational speed provides higher energy dissipation in the liquid phase over the disk surface. The specific dispersed power, [W/kg], was calculated according to the equation proposed by Moore [19]:

$$\varepsilon = \frac{1}{2 \cdot t_{res}} \left(\left(r_e^2 \cdot \omega^2 + {v_{r_e}}^2 \right) - \left(r_i^2 \cdot \omega^2 + {v_{r_i}}^2 \right) \right) \tag{2}$$

where t_{res} is the residence time of the liquid solution on the rotating disk between an external radius, r_e, where the fluid velocity is v_{re}, and an internal radius, r_i, where the fluid velocity is v_{ri}. The residence time can be calculated by the following relationship:

$$t_{res} = \frac{2 \cdot (r_e - r_i)}{v_{r_e} + v_{r_i}} \tag{3}$$

The values of the specific dispersion energy determined for a feeding point at 2 cm far from the disk center as function of the adopted values of rotational speed are reported in **Figure 4**. In this figure, the value of ε calculated for a feeding point 3 cm far from the disk center and at 147 rad/s is also reported. It is clear that increasing the disk speed and the distance of the feeding point from the center higher energy dissipation power occurs, producing better mixing conditions.

In conclusion, by increasing the local specific energy dispersion over the disk surface, micro-mixing at the contact point of the two reagent solutions is enhanced and HAP nanoparticles of smaller size are produced.

3.2. Production process of mg-doped HAP

It is well known that trace quantities of cations (i.e., Mg^{2+}, Zn^{2+}, Sr^{2+}) and/or anions (i.e., SiO_4^{4-}, F^-, CO_3^{2-}) in HAP play a pivotal role in its overall biological performances. Among substituting cations, magnesium is widely studied, being the fourth most abundant cation in the human body (0.44–1.23 wt%). Mg^{2+} substitution plays an essential role in the biologic environment due to its strong impact on the mineralization process, influencing both HAP crystal formation and growth [20], and increasing the HAP dissolution in human physiologic medium. In this work, the chance to produce Mg-doped nanoparticles by precipitation using

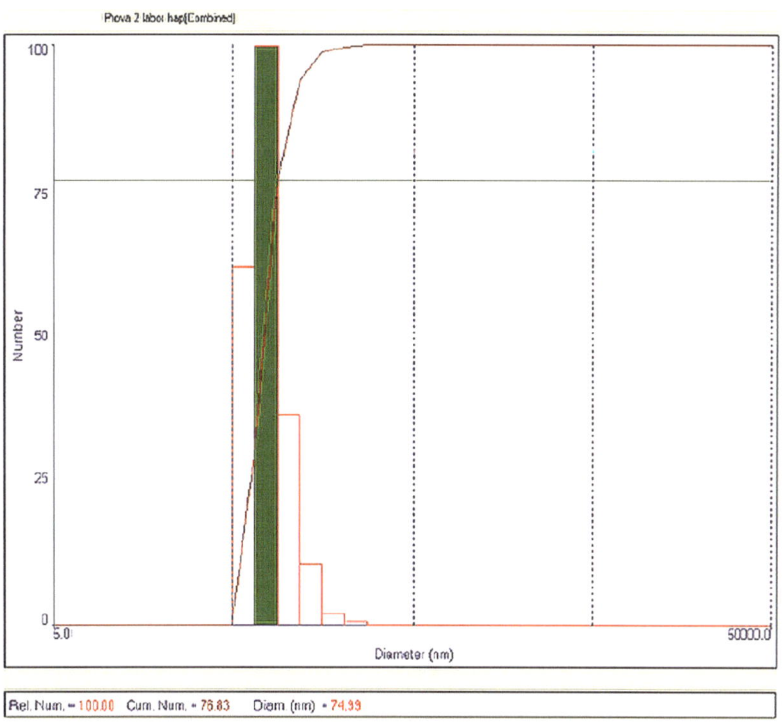

Figure 3. Size distribution measurement of HAP nanoparticles obtained with a rotational speed of 147 rad/s.

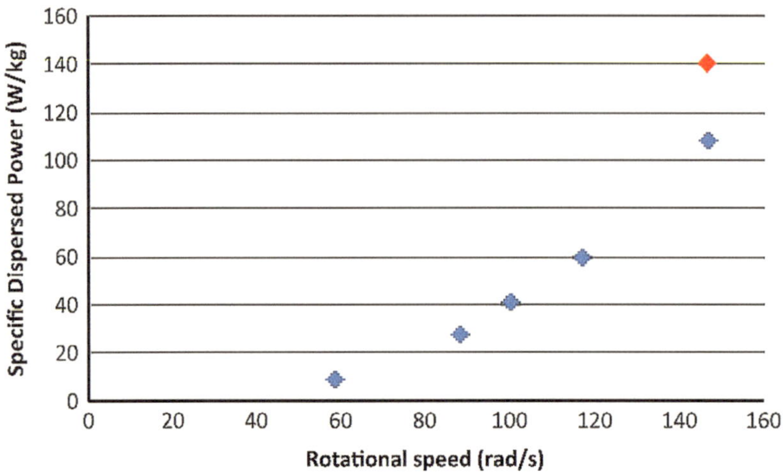

Figure 4. Specific dispersed power as a function of the rotational speed and reactant injection position. Blue points and red point refer to feed points at 2 and 3 cm from the disk center, respectively.

an SDR has been proven and the influence of the operating parameters on the nanoparticles size was investigated.

Mg-doped hydroxyapatite powders were prepared at room temperature (25°C) and air atmosphere. The precipitation reactor was performed by using the SDR 8.5 cm in diameter above described with a rotational velocity between 840 and 1500 rpm. The reaction took place among the following reagent aqueous solutions: ammonium phosphate dibasic ($(NH_4)_2HPO_4$ 0.264 M), ammonium hydroxide, (NH_4OH 2.853 M), calcium chloride ($CaCl_2$ 0.475 M) and magnesium chloride ($MgCl_2$ 0.028 M). The overall reaction was as follows:

$$(10-x)\ CaCl_2 + xMg\ Cl_2\ 6H_2O + 6(NH_4)_2\ HPO_4 + 8NH_4OH \rightarrow Ca_{10-x}\ Mg_x(PO_4)_6(OH)_2 + 20\ NH_4Cl + 6(1+x)\ H_2O \quad (4)$$

A value of x = 0.566 was chosen to determine a Mg^{2+}/Ca^{2+} molar ratio equal to 0.06, that is the value suggested by Landi et al. [21] to achieve the fastest bone growing rate. The NH_4OH solution was fed at the disk center, whereas the feed points of the two other reagent solutions were symmetrically located at 2 or 3 cm from the center of the disk. The details of the experimental work are reported elsewhere [22]. After each run, the obtained nanoparticles were first separated by the mother solution, then washed several times and dried for 96 h in a furnace at 80°C.

The size measurements of the produced particles were taken as above described. The X-ray diffraction (XRD) characterization of the HAP particles was performed using an XRD diffractometer (Philips PW1830 DY3558 Cu Kα, 40 kV, 30 mA). The analysis was made over a 2θ range of 2–70° at a scan rate of 0.5°C/min, with a sampling interval of 2.5 h. The crystallites average dimension was estimated from the X-ray diffractometer using the Scherrer's equation.

In addition, the B.E.T. surface area of the powder was measured by the Monosorb instrument supplied by Quantachrome. The adsorbed gas was N_2 (30%) and He (70%) at −196°C. The

http://dx.doi.org/10.5772/intechopen.71775

morphology of the powder was examined by scanning electron microscopy (SEM, 10 Auriga 405 Carl Zeiss). Infrared spectra of Mg-HAP powder were obtained using an infrared Fourier-transform spectrometer (FTIR, VERTEX 70 model by Bruker). Finally, in order to determine both HAP stoichiometry and the Mg^{2+}/Ca^{2+} molar ratio, chemical analysis of Mg^{2+} and Ca^{2+} was done by atomic absorption spectrophotometry (AAS, Agilent Technologies 200 series AA).

In the preliminary work, the effect of the rotational speed and the reagent flow rate on the nanoparticle size was noticed; thus, accordingly, two experimental work series were carried out to investigate the influence of these two operating variables. In **Tables 1** and **2** are reported the obtained results in terms of the nanoparticles size by changing the SD rotational speed and the overall flow rate of the reagent solutions, respectively.

The rotational speed has a very strong effect on the nanoparticle size (**Table 1**), as noticed in the experimental work on pure HAP, because of the influence of the hydrodynamics on the local micromixing. By feeding the reagent solutions at 2 cm from the disk center, comparable results are obtained at rotational speed equal to or higher than 1120 rpm because similar conditions of almost complete micromixing in the liquid over the SDR surface were attained. Moreover, from the results in **Table 1**, it is clear that the feed location largely affects the size of the produced particles. For a feed location at 3 cm from the disk center, at all the rotational speed, the size of the obtained particles was less than 100 nm and smaller than one of the particles obtained for the feed location at 2 cm from the disk center. In order to interpret the obtained results, it is useful to take into account that in any case the obtained particles are agglomeration of single nanoparticles, whose size is mainly affected by the local micromixing. The larger the micromixing, due to the increase in the rotational speed (**Table 1**), the smaller the single produced particles. The effect of crystal collision on the nanoparticle size is clearly shown by the results reported in **Table 2**, performed at constancy of rotational speed. By increasing the overall reagent solution flow rate, the residence time of the particle slurry suspension decreases and the probability of collisions as well. As a consequence, the smallest particle size is achieved at the maximum flow rate of 4 ml/s. In particular, by increasing eight times the overall feed flow rate from 0.5 up to 4 ml/s, the average size of the agglomerated particles decreases from 71 down to 52 nm.

The image of the nanoparticles obtained at the best operating conditions, that is, at 1400 rpm, location point 3 cm from the center and overall reagent solution flow rate of 4 ml/s, is reported in **Figure 5**.

SD rotational speed rate (rpm)	**SD rotational speed rate (rpm)**	
	Feed location at 2 cm	**Feed location at 3 cm**
840	392.8	97.0
980	305.1	71.7
1120	80.2	64.3
1260	75.2	56.9
1400	72.3	51.3

Table 1. Average particle size at different SD rotational speed (overall flow rate of the reagent solutions equal to 3 ml/s).

The average size and the standard deviations along the two geometrical axes are, respectively, 169 and 36 nm, along the major axis and 33 and 7 nm along the minor axis, and thus, the nanoparticles exhibit a length/width ratio around 5. The XRD pattern of the synthesized Mg-doped powder reported in **Figure 6** confirmed a nanocrystalline single-phase HAP and allowed

Overall liquid feed flow rate (ml/s)	Particle size (nm)	
	Average size	Standard deviation
0.5	70.7	5.4
1.0	58.8	7.4
2.0	51.9	6.8
3.0	51.3	7.8
4.0	51.3	7.8

Table 2. Nanoparticle size of mg-doped HAP at different values of the overall feed flow rates (feed location at 3 cm from the center and rotational speed equal to 1400 rpm).

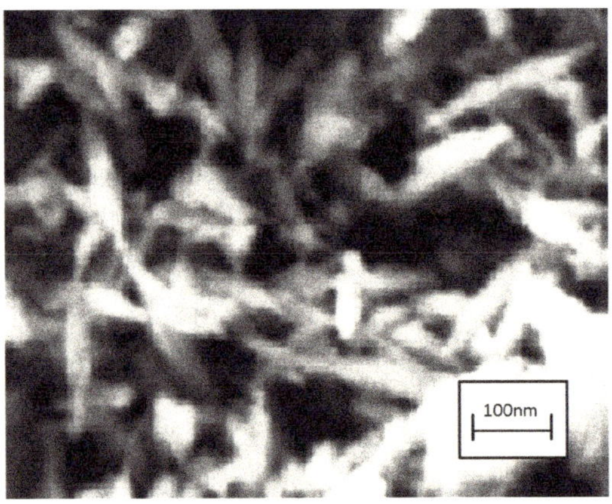

Figure 5. SEM image of the mg-doped HAP nanoparticles obtained at the best operating conditions.

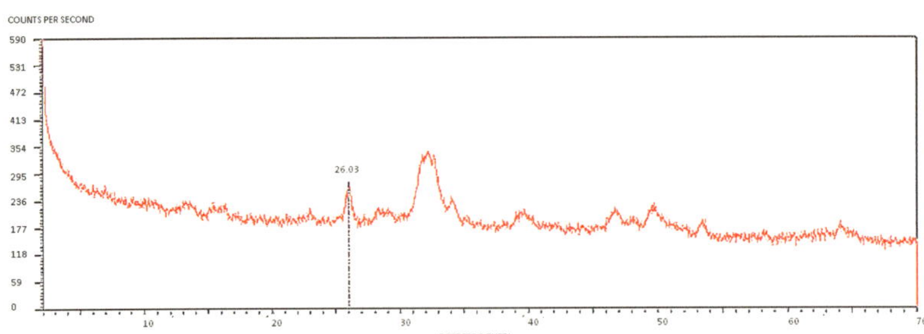

Figure 6. XRD pattern of the produced mg-HAP powder.

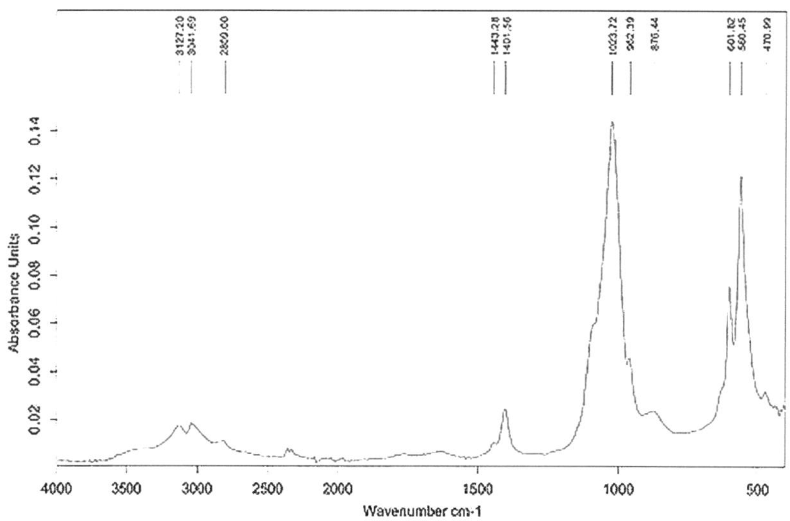

Figure 7. FT-IR spectra of the produced mg-HAP sample.

to estimate crystallite dimension of 5 nm by applying the Scherrer's formula. The estimated size of the crystallites is consistent with the hypothesis of agglomeration for the produced particles.

The B.E.T. results gave a specific area of 132.6 m^2/g, which is higher than the values reported by Landi [21], in the range 90–125 m^2/g. The FT-IR spectra reported in **Figure 7** show the typical phosphate bands of hydroxyapatite compounds located at 980–1100 cm^{-1} (asymmetric stretching) and at 560–600 cm^{-1} (asymmetric bending).

Finally, from the Mg-HAP analysis by atomic adsorption spectrophotometry a molar ratio Mg^{2+}/Ca^{2+} equal to 0.06 resulted, in according with the target fixed in this work.

4. CFD modeling of the SDR performance in HAP synthesis

The progressive progress in computational fluid dynamics (CFD) techniques and in the available computing power encourages the application of this modeling approach across multiple engineering fields and, in particular, in the area of chemical reaction engineering. However, in the case of spinning disk reactors applied to the synthesis of nanoparticles, a few of CFD studies are present in literature, in spite of the expected benefits for the physical interpretation of the occurring physical–chemical phenomena. In this section, a careful attempt has been made to model the hydrodynamics of the three-phase system over the disk surface and to interpret the obtained experimental results with reference to the HAP nanoparticles production.

SDR performances are strongly affected by the adopted operating conditions. Consequently, a fine description of thin-film hydrodynamics appears as essential in studying and optimizing the operating conditions of an SDR.

The use of an SDR should promote the mixing among the reagents, which leads to very low micromixing time, less than 1 ms, to maximize the reagent concentrations and one of the products. In case of no adequate mixing, in fact, the two reagent solution streams may give rise to a wide segregate area where one of the reagents predominates. Once this segregation is established, the reaction occurs mainly outside the selected pH value, producing the formation of larger particles which tend to agglomerate away from their feed point because of the high residual product concentration.

To avoid this unfavorable situation, the optimal condition should be set, promoting a rapid mixing of the reagent streams, maximizing the reaction rate between the reagents, where high nucleation rate occurs, and minimizing the residual supersaturation available for the growth of the formed particle.

These conditions are all favored by a high velocity field in the overall film thickness developed on the disk. Some simplified descriptions of the film hydrodynamics were proposed in literature, starting from the simplified Nusselt model, the Pigford, Lepehin and Riabchuk models, just to mention a few. As matter of fact, Burns et al. [9] reported that the Nusselt model returns a rough description of the film thickness at high Ekman numbers, with an average overprediction around 10%. Based on the experimental results reported in this Burns' work, Bhatelia et al. [10] implemented and validated a CFD model where no specific liquid film turbulence was addressed, producing results not enough accurate to grasp the true film hydraulics.

The approach here proposed to simulate the behavior of the reaction precipitation system over the adopted SDR moves from a CFD hydrodynamic model developed by de Caprariis et al. [11]. In this study, the CFD model is extended to the prediction of the average size of the produced HAP nanoparticles. The hydroxyapatite production was studied with the aim to predict the SDR performances at various operating conditions. A nanoparticle diameter estimation to be compared with the available experimental data was derived by the implementation of a population balance equation (PBE).

As previously described, the production of nanoparticles of hydroxyapatite by chemical precipitation reaction took place in an SDR operating in continuous mode and consisting of an inner rotating disk 8.5 cm in diameter. Three reagent solutions were injected onto the disk at three selected feed points: the ammonium hydroxide at the distance of 1 mm from the disk and the other two reagents at 2 cm from the disk center, in opposite positions.

The first step of the simulation procedure consisted of the generation of a stable, stationary and continuous liquid film onto the disk surface. This initial condition was addressed by feeding only the NaOH water solution (10%wt) at the rate of 80 ml/min. Once a stationary liquid phase was established, the two reagents $CaCl_2$ and $(NH_4)_2HPO_4$ were continuously fed in the form of water solution (5.6 and 3.5%wt, respectively), both at the same flow rate of 100 ml/min, at a position of 2 cm from the disk center. A calcium/phosphate (Ca/P) stoichiometric ratio of 1.67 was assumed. The rotational velocity was fixed at 146.5 rad/s. The reaction takes place between calcium chloride and ammonium phosphate, in the presence of ammonium hydroxide, according to the stoichiometry described by Eq. (1).

On the basis of the disk geometry, a computational grid necessary to resolve the CFD model was built in the Gambit environment. The computational domain has considered only the zone of the disk from a radius of 2 cm ahead, that is where the liquid height is approximately constant and the reaction takes place. In fact, according to the liquid profiles reported in

Figure 8, in the examined domain the film thickness is quite constant and it is the only one where the reaction occurs, since the reagents are injected 2 cm away from center of the disk. Therefore, the adopted grid for the simulation work consists of a cylinder 8.5 cm in diameter and 70 μm in height, composed by 260,000 cells. The mesh is structured and made of hexahedral cells, so that the flux is orthogonal to the faces of each cell in the radial direction, limiting the numerical diffusion errors especially in the presence of convective fluxes.

The numerical simulations were performed in the ANSYS Fluent v.14.5 environment, a commercial CFD package based on finite volume resolution method. The reaction was modeled according to the Eulerian multiphase model. This model was chosen because the length of the interface between the forming crystals and the liquid medium is by far shorter than the characteristic length of the computational domain. In the framework of this model, the physical system was simulated by a continuous liquid phase containing all the reacting species and a solid phase consisting of the nucleating hydroxyapatite nanoparticles formed by the reaction. The Eulerian model solves a set of *n* continuity and momentum equations for each phase, coupled through the interphase and pressure exchange coefficients. The turbulence was modeled according to *k*-epsilon model.

The precipitation reaction was described according to the finite rate model implemented with literature kinetic data [23]. The set of equations consists of the balances of momentum, mass, energy and solid particle population. The population balance equation (PBE) aiming to predict the size distribution of the particles is written in terms of density function $n(V,t)$ as follows:

$$\frac{\partial}{\partial t}[n(V,t)] + \nabla\left[\vec{u}\, n(V,t)\right] + G = A_B + A_D + B_B + B_D \qquad (5)$$

where $\dot{n_V}$ is the nucleation rate (#/m^{3}s). The PBE can be solved once the boundary and initial conditions are set:

- BC: $n(V=0,t) = \dot{n_0}$
- IC: $n(V,t=0) = n_V$

In the PBE, G is the growth term, A_B and A_D are the birth and death rate due to aggregation terms, respectively, and B_B and B_D are the birth and death rate due to breakage terms, respectively. In the considered process, however, all the terms, apart from the nucleation

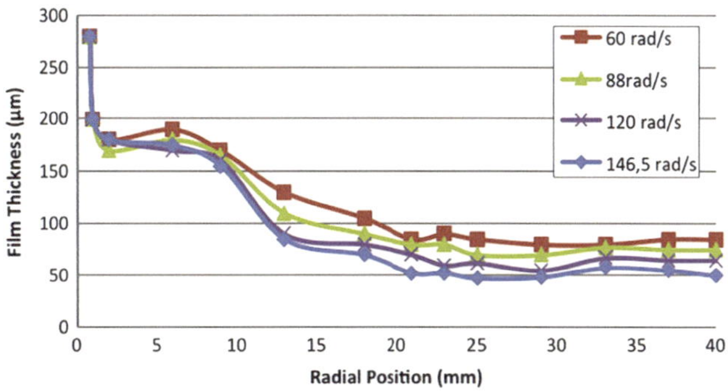

Figure 8. Film thickness profile [10].

contribution, have been ignored due the particular characteristics of the reaction. In fact, when the precipitation reaction takes place at complete micromixing conditions, it can be considered that most of the supersaturation ratio, that is, the driving force, is consumed by the nucleation, leaving only a reduced driving force for the remaining phenomena, that is, solid growth and aggregation. A constant nucleation rate $\dot{n_V} = 10^{11}$ #/m^{3}s, equal to an average value derived from the literature data for this class of reactions, was assumed.

The PBE can be solved according to different approaches, and in this chapter, the Quadrature Methods of Moments (QMOM) was adopted [24]. This model allows the calculations of the moments describing the population balance through a relatively reduced set of equations, furthermore limiting the computational errors. The main advantages of this approach are to involve few variables (from six to eight different moments) and to allow the dynamic calculation of the size bins, obtained however at a quite high computational cost. Further details about the model choice and its setting are reported by Dugo [25]. The main results of the CFD simulation are briefly shown and discussed below.

The tangential velocity profiles of the liquid phase at the disk surface and at the liquid film maximum height resulted by the CFD simulation are reported in **Figure 9**. The liquid velocity continuously increases from the center to the periphery of the disk, as expected.

The contours of the concentrations of the two reagent streams computed at the middle of the film (35 μm) are reported in **Figure 10**. These contours show that the maximum concentration of each reagent at the feeding point progressively lows down along the disk, due to the reaction occurrence. It has to be noticed that from the quantitative point of view the complete mixing gives rise along the disk to an average stoichiometric ratio between the two reagents of approximately 1.67.

The precipitation reaction is nearly instantaneous and starts as soon as the reagent streams get in contact. Hence, the maximum reaction rate is located at the contact points, as clearly shown in **Figure 11**.

The HAP mass fraction contour showed in left side of **Figure 11** confirms that the HAP production starting point occurs at the feed point of the two reagents, where the calculated reaction rate shows the highest values (right side of **Figure 11**). From these points, ahead the

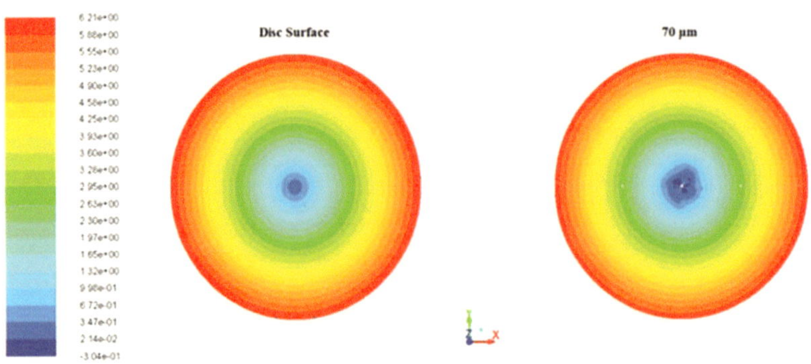

Figure 9. Tangential velocity profiles of the liquid phase at the disk surface and at the maximum film height (70 μm).

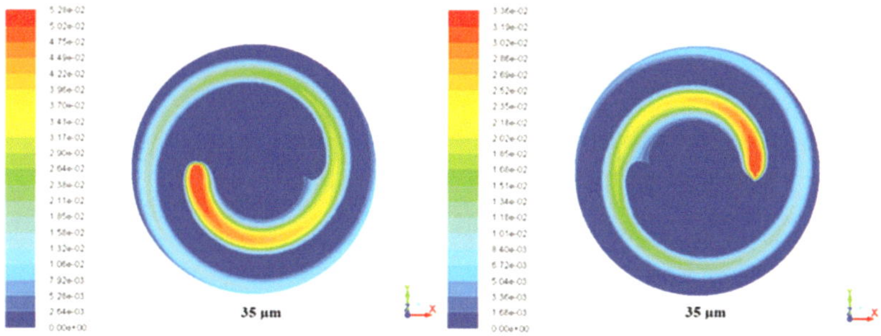

Figure 10. Reagent concentration contours at a film height of 35 μm: $CaCl_2$ left, $(NH_4)_2HPO_4$ right.

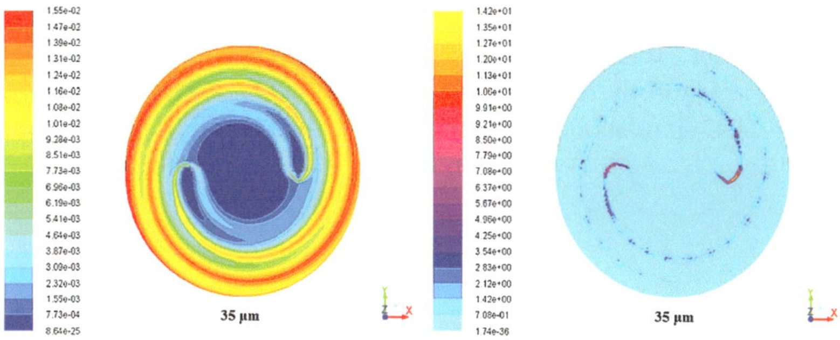

Figure 11. Hydroxyapatite concentration contour in the liquid phase at a film height of 35 μm, left. Reaction rate contour, right.

HAP concentration increases for the advances of the reaction along the disk until a maximum concentration at the disk edge.

On the basis of the calculated mass produced by the reaction, the HAP crystallite dimension was predicted by means of the PBE as here after detailed.

The results of concern of the PBE solution are the moments of different orders that are directly related to the particle diameters. In particular, the moments of order from 0 to 3 are linked to the representative diameters, d_{10} and d_{32}, by Eqs. (6) and (7), respectively:

$$d_{10} = \frac{m_1}{m_0} \tag{6}$$

$$d_{32} = \frac{m_3}{m_2} \tag{7}$$

Because the estimation of the d_{32} diameter is based on the m_2 and m_3 moments, linked to the surface and volume shape factors, respectively, the corresponding figure can be considered more accurate in inferring the true particle size.

The values of the moments calculated at the disk edge are shown in **Table 3**, allowing the estimation of the formed nanoparticle size in the range d_{10} = 2.19 nm and d_{32} = 4.8 nm. These results should be considered as referring to the dimension of the crystallite, that is, the crystals

Moments	
m_0	4.12×10^{7}
m_1	9.04×10^{-2}
m_2	1.63×10^{-11}
m_3	7.82×10^{-20}

Table 3. Values of the moments obtained from the simulations.

born due to the nucleation phenomenon, since no crystal growth and agglomeration phenomena were taken into account in the simulations.

The dimension of the crystallite of hydroxyapatite was measured in previous works [13, 22] with X-ray diffraction technique, calculating the value of the diameter by the Scherrer's formula. The images of the HPA crystals reported in Section 3 and considered as agglomerated crystals are consistent, and the estimated crystallite diameter is about 5 nm. The d_{32} value obtained from the CFD simulation proves, thus, the reliability of the developed model.

5. Conclusions

The biocompatible characteristics of hydroxyapatite are emphasized when its mass has a very high specific surface, as in case of nanoparticles. In this chapter, it has shown that the HAP production process can be effectively performed by chemical precipitation by using a spinning disk reactor. By operating at a disk rotational speed of 1500 rpm, pure HAP nanoparticles around 70 nm are obtained. When the reaction is operated in the presence of $MgCl_2$, Mg-doped HAP nanoparticles are obtained down to 51 nm in size. In this case, the adopted analytical techniques ascertained both the nature of HAP and a molar ratio Mg^{2+}/Ca^{2+} equal to 0.06, needed to achieve the fastest bone growing rate. The results in terms of the obtained nanoparticle sizes are worthwhile because they refer to a production process which can be carried out in continuous mode, whereas most of the results in literature concern preparation in batch mode. In this chapter, the effects of the disk rotating speed and the feed location of the reagent solutions on the produced particle size were clearly shown and discussed. Then, a CFD model was developed in order to describe the hydrodynamics and the reaction-precipitation process in the film thickness formed over the SDR surface. The interest in developing such a tool relies in the possibility to predict the outcome of reagent mixing and chemical reaction processes into the system domain, a prerequisite to estimate the particle size distribution of the product obtained by chemical precipitation.

The results show that the SDR is an effective device in performing this class of reactions where the mixing of the reagents is of fundamental importance. Hydroxyapatite, indeed, is produced in the liquid phase instantaneously as soon as the reagents enter in contact. The population balance equation added to the hydrodynamic model allows an estimation of the particle diameters. Because only the nucleation was taken into account in the PBE equation, the prediction concerned only the size of the HAP crystallite, which results of 4.3 nm. This value is in a good agreement with literature experimental data.

http://dx.doi.org/10.5772/intechopen.71775

Author details

Benedetta de Caprariis, Angelo Chianese, Marco Stoller and Nicola Verdone*

*Address all correspondence to: nicola.verdone@uniroma1.it

Department of Chemical, Material, Environmental Engineering, Sapienza University of Rome, Rome, Italy

References

[1] Paz A, Guadarrama D, López M, González JE, Brizuela N, Aragón J. A comparative study of hydroxyapatite nanoparticles synthesized by different routes. Quimica Nova. 2012;**35**:1724-1727

[2] Hung LH, Lee AP. Microfluidic devices for the synthesis of nanoparticles and biomaterials. Journal of Medical and Biological Engineering. 2007;**27**:1-6

[3] Cafiero LM, Baffi G, Chianese A, Jachuck RJJ. Process intensification: Precipitation of barium sulphate using a spinning disc reactor (SDR). In: Proceedings of 14th European Symposium on Ind. Crystallization; 12–16-09-1999; Cambridge

[4] Baffi G, Cafiero ML, Chianese A, Jachuck RJJ. Process intensification: Precipitation of barium sulphate using a spinning disc reactor (SDR). Industrial and Engineering Chemistry Research. 2002;**41**:5240-5246

[5] Trippa G, Hetherington P, Jachuck R. Process intensification: Precipitation of calcium carbonate from the carbonation reaction of lime water using a spinning disc reactor. In: Proceedings of 15th International Symposium on Industrial Crystallization; 15-18-09-2002; Sorrento, Italy

[6] Raston CL, Anantachoke N, Makha M, Reutrakul V, Smith NC, Saunders M. Fine tuning the production of nanosized b-carotene particles using spinning disc processing. Journal of the American Chemical Society. 2006;**128**:13847-13853

[7] Loh JW, Schneider J, Carter M, Saunders M, Lim L. Spinning disc processing technology: Potential for large-scale manufacture of chitosan nanoparticles. Journal of Pharmaceutical Sciences. 2010;**99**:4326-4336

[8] Dabir H, Davarpanah M, Ahmadpour A. Effects of different operating parameters on the particle size of silver chloride nanoparticles prepared in a spinning disc reactor. Applied Physics A. 2015;**120**:105-113

[9] Burns JR, Ramshav C, Jachuck RJ. Measurement of liquid film thickness and the determination of spin-up radius on a rotating disc using an electrical resistance technique. Chemical Engineering Science. 2015;**58**:2245-2253

[10] Bhatelia TJ, Utikar RP, Pareek VK, Tade MO. Characterizing liquid film thickness in spinning disc reactors. Proceedings of the 7th International Conference on CFD in the Minerals and Process Industries; 9–11-12-2009; Melbourne, Australia

[11] de Caprariis B, Di Rita M, Stoller M, Verdone N, Chianese A. Reaction-precipitation by a spinning disc reactor: Influence of hydrodynamics on nanoparticles production. Chemical Engineering Science. 2012;**76**:73-80

[12] Plasari E, Muhr H, Rousseaux JM, Vial C. CFD simulation of precipitation in the sliding-surface mixing device. Chemical Engineering Science. 2001;**56**:1677-1685

[13] de Caprariis B, Stoller M, Chianese A, Verdone N. CFD model of a spinning disc reactor for nanoparticle production. Chemical Engineering Transactions. 2015;**43**:757-761

[14] Hounslow MJ, Mumtaz HS. Aggregation during precipitation from solution: An experimental investigation using a Poiseuille flow. Chemical Engineering Science. 2000;**55**:5671-5681

[15] Parisi M, Stoller M, Chianese A. Production of nanoparticles of hydroxyapatite by using a spinning disc reactor. Chemical Engineering Transactions. 2011;**24**:211-217

[16] Stoller M, Miranda L, Chianese A. Optimal feed location in a spinning disc reactor for the production of TiO_2 nanoparticles. Chemical Engineering Transactions. 2009;**17**:993-999

[17] Martins MA, Santos C, Almeida MM, Costa MEV. Hydroxyapatite micro- and nanoparticles: Nucleation and growth mechanisms in the presence of citrate species. Journal of Colloid and Interface Science. 2008;**318**:210-216

[18] Pang YX, Bao X. Influence of temperature, ripening time and calcination on the morphology and crystallinity of hydroxyapatite nanoparticles. Journal of the European Ceramic Society. 2003;**23**:1697-1704

[19] Moore SR. Mass Transfer to Thin Liquid on Rotating Surfaces with and without Chemical Reaction [Thesis]. University of Newcastle upon Tyne; 1986

[20] Pina S, Olhero S, Gheduzzi S, Miles A, Ferreira J. Influence of setting liquid composition and liquid-to-powder ratio on properties of a mg-substituted calcium phosphate cement. Acta Biomaterialia. 2009;**5**:1233-1240

[21] Landi E, Logroscino G, Proietti L, Tampieri A, Sandri M, Sipro S. Biomimetic mg-substituted hydroxyapatite: From synthesis to in vivo behavior. Journal of Materials Science. Materials in Medicine. 2008;**19**:239-247

[22] D'Intino AF, de Caprariis B, Santarelli ML, Verdone N, Chianese A. Best operating conditions to produce hydroxyapatite nanoparticles by means of a spinning disc reactor. Frontiers of Chemical Science and Engineering. 2014;**8**:156-160

[23] Liu C, Huang Y, Shen W, Cui J. Kinetics of hydroxyapatite precipitation at pH 10 to 11. Biomaterials. 2001;**22**:301-306

[24] Marchisio DL, Virgil RD, Fox RO. Quadrature method of moments for aggregation-breakage processes. Journal of Colloid and Interface Science. 2003;**258**:322-334

[25] Dugo F. Modellizzazione fluidodinamica di un reattore a disco rotante per la produzione di nanoparticelle di idrossiapatite [Master Thesis]. Library of the Chemical Engineering Department. Sapienza University of Rome; 2014

Chapter 3

HAp Nanofibers Grown with Crystalline Preferential Orientation and Its Influence in Mechanical Properties of Organic-Inorganic Composite Materials

Eric M. Rivera-Muñoz, Rodrigo Velázquez-Castillo, Susana Alonso-Sierra, J. Rafael Alanís-Gómez, Beatriz Millán-Malo, Lauro Bucio-Galindo, Rafael Huirache-Acuña, Alejandro Manzano-Ramírez, Rufino Nava and Miguel Apátiga-Castro

Additional information is available at the end of the chapter

http://dx.doi.org/10.5772/intechopen.71850

Abstract

There are several synthesis techniques to obtain hydroxyapatite (HAp). Some use surfactant agents, amino acids or halogen salts to control structural nucleation and crystal growth. In others, the use of hydrothermal process to carry out the reaction is effective for HAp synthesis. Microwave-assisted hydrothermal method (MAHM) has been successfully applied in the synthesis of HAp nanostructures, which present well-defined morphologies, high crystallinity and high purity. This is important because nano-HAp is attracting interest as a biomaterial for use in prosthetic applications due to its similarity in size, crystallinity and chemical composition with human hard tissue. In this chapter, developments in obtaining HAp nanofibers, with a crystal growth with preferential orientation, as well as morphology control achieved by using the MAHM is discussed. Also, the synthesized fibers were used to cast ceramics with controlled and interconnected porosity through the modified gelcasting process. Then, these HAp ceramics were impregnated with a water solution of gelatin in order to obtain an organic-inorganic composite material, similar to natural bone tissue. The maximum compressive strengths were determined and the composite materials showed mechanical properties that make them suitable to be used as bone tissue implants.

Keywords: bottom-up synthesis, HAp nanofibers, mechanical properties, microwave-assisted hydrothermal method, biomimetic material

1. Introduction

Hydroxyapatite (HAp) is an inorganic compound made of calcium phosphate and hydroxyl groups, with a stoichiometry of ($Ca_{10}(PO_4)_6(OH)_2$) and a hexagonal crystal structure. Natural HAp is the main mineral constituent of hard human tissues such bones and teeth, and it has the same crystal structure, but a stoichiometry slightly different from the synthetic one since some calcium ions can be exchanged by other metallic ions, such magnesium or sodium. The production of synthetic HAp has acquired a great significance in recent years due to its excellent properties of biocompatibility, bioactivity, osteoconductivity and osteoinductivity. Those properties are produced, in part, by its similarity in structure and chemical composition to those in the natural HAp [1]. Consequently, synthetic HAp has been widely applied in health-related fields as a replace, bone reconstructing or implant material in the restoration of damaged bones or teeth [2].

In spite of HAp is mainly useful as a biomaterial for implants in bone tissue, other interesting application have been developed in recent years. Some authors have explored the use of the HAp piezoelectric properties [3–5], some HAp-collagen composite materials have been studied to be applied in bone regeneration and osteogenesis using their piezoelectric properties [6, 7] or other composite materials based on HAp use this property to be used as a immunosensor material [8]. In addition, this HAp property was also studied to analyze its impact in the design and development of ferroelectric memories and micro-electro-mechanical systems, and consequently, some materials have been obtained for those applications [9]. Other interesting uses of HAp have been reported as catalysts [10–12], especially for the synthesis of organic compounds [13, 14], and some application in obtaining ecological biofuels [15, 16]. Additionally, the HAp has been used in water treatments [17–19] and for the environment remediation [20–22] to remove heavy metals. All these diverse applications have made the HAp a quite interesting material and the relevance of its synthesis and production methods are still increasing.

2. Different hydroxyapatite synthesis methods

Several synthesis methods have been proposed by many scientists in order to obtain HAp. The morphology, dimensions, crystalline quality and purity of the synthesized HAp have an impact on its properties, and the properties determine the final application of the HAp; for that reason, those structural characteristics often tried to control the synthesis process. **Table 1** summarizes the different HAp synthesis methods, their reaction times and the corresponding obtained morphology.

Sol-gel is one of the most used methods to obtain HAp. Here, the chemical substances, used as precursors, are dissolved in water (wet chemistry), and then this solution are blended with some monomer molecules to form a sol (colloid) at the beginning of the reaction, and posteriorly, the

Type of reaction	Reaction time	Reaction features	Reaction temperature	Obtained morphology
Sol-gel	24–120 h	A little energy available inside the reactor	25–45°C	Nanoparticles nanorods
Solid-state reaction	Several hours	Diffusion difficulties, calcination process required	25–1250°C	Nanoparticles
Conventional hydrothermal	Several minutes	Inhomogeneous temperature inside reactor	170°C	Nanoparticles, nanowires, nanofibers and nanoplates

Table 1. Different HAp synthesis methods and their corresponding outcome.

polymerization process continues until the formation of the gel. The chemical reaction takes place at room temperature, and therefore, there is little amount of energy for the reaction and, consequently, it takes several days [23–25], and the morphology mainly produce are particles.

An interesting example of wet chemical reaction is obtaining HAp from eggshells. The hen eggshells are made basically of calcium carbonate. In the synthesis to obtain HAp, the egg-shells are cleaned and milled; posteriorly, the powder undergoes an initial thermal treatment at 450°C to eliminate organic residues and after that, the calcination of the powder is made to transform the calcium carbonate into calcium oxide. This calcium oxide could be hydrated to form calcium hydroxide, which is put in reaction with a solution of phosphoric acid to finally obtain the HAp [26]. In a variation of this reaction, the calcium oxide could be put in contact with a water solution of calcium phosphate in order to obtain the HAp [27].

Other authors have used the calcium oxide to obtain HAp through a solid-state reaction. The calcium oxide is mixed with phosphorus oxide and other additives; then, the mixture is sintered inside an oven at 1250°C, and the HAp is finally obtained [28]. In other solid-state method to produce HAp, diammonium phosphate, sodium bicarbonate and calcium nitrate are used as the main precursors of the synthesis. The reactants are mixed in the proportion to achieve the stoichiometric of HAp and milled. The mixture is aged at room temperature and posteriorly, this blend is washed, dried and calcined to produce HAp [29]. In several cases, where a solid-state method was used to obtain the HAp, the final product was mixed with other by-products.

Although the aforementioned synthesis methods are important for the HAp obtention, in most of the works where those procedures were used the authors did not report a control on the morphology, size, crystalline quality or structural parameters of the HAp aggregates, which made suppose that those structural features were difficult to regulate through those methods.

In other more innovative synthesis methods, the production of HAp involves the use of different chemical substances such surfactants, amino acids or halogen salts to control the nucleation and crystal growth. The last one has, as a consequence, the control on the final morphology of HAp assembly [30–33]. Additionally, some of those methods use a conventional hydrothermal process (CHP) to carry out the synthesis reaction. In CHP, the precursors are dissolved in water and put inside an autoclave. The heat to perform the reaction comes from the outer side of the reactor walls, and produces a temperature gradient in the solution. Mechanical agitation can be suitable to help in the heat distribution, and thus, try to generate a homogeneous temperature in the whole reactor. The steam produced by water increases the pressure on the solution and therefore, the boiling point of water is increased and the solution can reach a higher temperature. This higher temperature in the reactor can increase the reaction rate. A measured supply of heat makes possible to have a control on the temperature of the reacting solution, and thus, also a control on the pressure. These reaction conditions make the crystallizing substances possible in a short time; additionally, the size and crystal quality can be controlled during the crystal growth.

The CHP has demonstrated to be effective for the HAp synthesis; the use of this method has been improved due to the relative low cost and simplicity to synthesize HAp in a large scale and with a high purity [34–37]. The CHP experienced an important development when a microwave oven was used for the synthesis reaction. The microwaves provide the energy required to carry out a chemical reaction. The emission of microwaves could be focused into the reactor to reduce the amount of energy used. Thus, the microwave-assisted hydrothermal method (MAHM) was developed. Currently, this method is applied in the crystal growth of multiple substances with an excellent control on their morphology, size and preferential crystal orientation. Several works have reported the use of MAHM in the synthesis of HAp nanostructures, which possess well-defined morphologies, high crystallinity and purity [38–44]. According to literature, the structures more often synthesized by this synthesis method are nanoparticles, nanorods and nanofibers.

3. The microwave-assisted hydrothermal method

In the MAHM, the electric field of microwaves interacts with the dipole moment of molecules to produce heat. When a molecule is irradiated with a microwave, its dipole moment tries to align with the electric field vector. In an electromagnetic wave, the electric field is oscillating, and the dipole moment vector always attempts to be aligned with it. During the aligning process, the molecules disperse energy in the form of heat, which is produced by intermolecular frictions, collisions and dielectric losses. If the dipole moment of the molecules in a sample has not enough time to align with the electric field or if their reorientation occurs in a short time, the sample is not warming up.

In a conventional microwave oven, the magnetron produces electromagnetic wave with a frequency of 2.45 GHz and the waves possess energy of 0.0016 eV, but the electric field spins 4.9×10^9 times in a second. The dipole moment in the molecules has enough time to align with the electric field, but they cannot oscillate with the same frequency. Therefore, phase differences between the electric field and the dipole moments are generated, and these phase differences cause energy dispersion, and the lost energy has the form of heat, which is generated by the intermolecular collisions, frictions and dielectric losses. The heat produced increases the temperature of the sample. The presence of ions could accelerate the heating process. In that way, the reacting mixture inside the reactor could be heated quite homogeneously in a microwave oven. A better heat distribution produces a quite similar temperature into the whole reactor, and this cause that the chemical reaction take place with a similar rate everywhere into the reactor; thus, the crystal growth is rather homogeneous and, consequently, the size and the crystalline quality in the obtained crystal can be controlled.

Some materials can reflect the microwaves, for example, metals, and they do not allow the wave pass. In other materials, the microwaves produce low dielectric losses and then, those materials such glass, Teflon and quartz are transparent and they permit the microwave to pass. In MAHM, the vessels (reactors) are made of transparent materials. The absorbing materials are dielectrics and they have great dielectric losses generated by the interactions of microwaves; therefore, they can be heated by microwaves.

In this work, a detailed description is made of how the microwave-assisted hydrothermal method was used to prepare several types of HAp morphologies, from particles until nanofibers with a preferential crystalline orientation in the [300]. Some of these nanostructures had a high crystallinity and good purity, which were a consequence of the synthesis method used. A careful control on the operation conditions in the microwave oven concerning pressure and temperature allowed obtaining controlled heating and cooling rates to produce the HAp. The suitable chemical composition in the reacting mixture, and posteriorly, the use of the glutamic acid (GA) as the substance to guide and control the crystal growth of HAp nanostructures, were the main factors to obtain the different morphologies in the HAp assemblies.

Posteriorly, the HAp nanofibers were selected to produce a ceramic with a controlled porosity through the modified gelcasting process (MGCP). Porosity is rather relevant for the HAp in its application as a bone tissue implant. Dimension of pores must be large enough to permit the flow of nutrients, substances and cells with the purpose of carry out the regeneration process in the damaged bone tissue.

Once the porous ceramic was made, an organic phase made of protein was added in order to obtain an inorganic-organic composite material. The chemical interaction between both phases produced synergic mechanical properties.

4. Synthesis of HAp with different morphologies through MAHM

The long experimental way to obtain different morphologies in the HAp began with the formulation of the reacting mixture to perform the synthesis reactions. The chemical substances chosen for the synthesis were calcium nitrate [$Ca(NO_3)_2 \cdot 4H_2O$] as the source of calcium, monobasic potassium phosphate [KH_2PO_4] as the source of phosphate groups and potassium hydroxide [KOH] to supply the hydroxyl groups. These inorganic compounds were dissolved separately in 100 ml of distillated water to achieve the stoichiometry of HAp. After that, the solutions were mixed and a 400-ml new dissolution (reacting mixture) was made, which was put inside eight Teflon tubes (50 ml each). The tubes were placed inside the microwave oven. This reaction was numerated as "1" (**Table 2**), and took place within the microwave oven at 150°C at a pressure of 690 kPa, 100 W power was used for a reaction time of 90 min. When the reaction was over, the final product was washed with ethanol, filtered and dried at room temperature. The HAp synthesized was analyzed by X-ray diffraction (XRD) by powders to identify the crystalline phases present in the sample and the respective diffractogram is shown in **Figure 1**(**a**). The phase identification was done by comparison using the powder diffraction file (PDF) data bank provided by the International Center for Diffraction Data (ICDD), and only one crystalline phase was found, which correspond to HAp, with PDF #86-1199. The presence of a single crystalline phase indicates that the synthesis method is suitable to

Reaction	$Ca(NO_3)_2$	Glutamic acid	HAp crystalline features	Obtained morphology
1	CNC	0.0	Good purity and satisfactory crystalline quality	Nanoparticles
2	CNC	0.2 (CNC)	Better crystalline quality and defined morphology	Nanowires-nanofibers
3	CNC	0.9 (CNC)	Excellent crystalline quality and preferential crystalline orientation	Nanofibers
4	CNC	2 (CNC)	Remarkable preferential crystalline orientation and structural organization	Nanofiber closely packed in microfibers
5	CNC	2.5 (CNC)	Decrease in the preferential crystalline orientation and a multilayer structure	Nanoplates

Table 2. Different HAp morphologies obtained by MAHM.

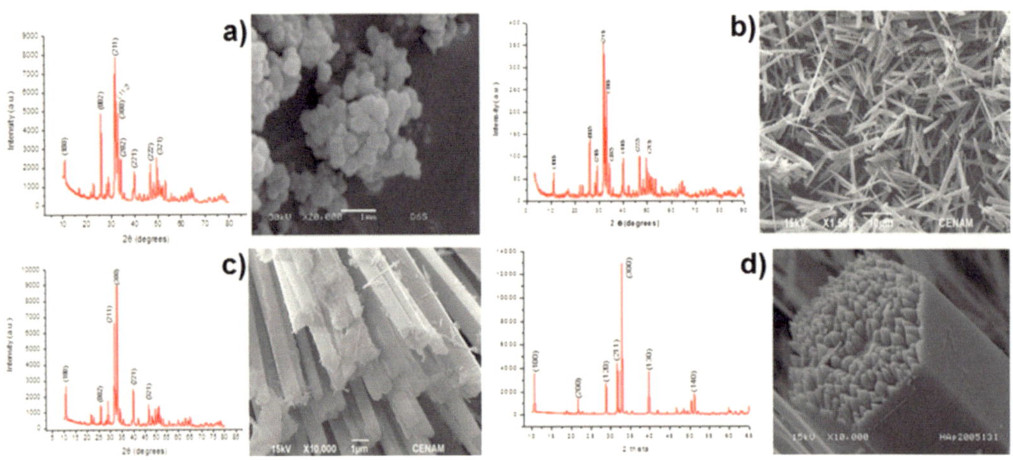

Figure 1. X-ray diffractograms and morphology of the HAp structures synthesized by the MAHM in reactions 1 (a), 2 (b), 3 (c) and 4 (d).

obtain HAp with a good purity. The diffractogram shows a background with a high noise level and the Bragg reflection are well-enough defined, those are indicative of an acceptable crystalline quality. Afterward, the HAp sample was observed by means of a scanning electron microscope, and the morphology found correspond to particles, as can be seen in **Figure 1**(**a**). The diameter in the particles was in the range from 90 until 160 nm. This reaction was useful to probe that HAp can be synthesized through the MAHM.

The next step was to include the glutamic acid (GA) in the chemical formulation of the reacting mixture and analyze the influence of this amino acid in the morphology of the HAp. The concentration of the other components in the reacting mixture was kept constant and the amount of GA varied. The reaction conditions in the microwave oven were the same as those described for reaction 1.

The GA has two carboxyl groups and one amine group. The nitrogen and the oxygen in these functional groups possess electron-pairs, which can be donated to the calcium ions in their empty "d" orbitals in order to form coordinate covalent bonds. Thus, the GA molecules bond the calcium ions and the acid can guide the crystal growth of HAp.

In reaction 2, the GA was added to the reacting mixture. The calcium nitrate concentration (CNC) was the reference and the GA concentration was 0.2 CNC in this reaction. The GA was dissolved in 200 ml of distilled water along with the calcium nitrate. At first, the solution was cloudy due to the GA solubility and for this reason, mechanical agitation was performed for about 2 h; posteriorly, the solution turned transparent and both chemical substances became dissolved. It is possible that the coordinate covalent bonds between Ca ions and the oxygen or nitrogen atoms could occur during this solution preparation, and consequently, the solubility of both substances was facilitated by these bonds formations.

The other reacting substances, KH_2PO_4 and KOH, were dissolved together in 200 ml of distilled water to form an additional solution. This solution was agitated for 15 min. Posteriorly, both dissolutions were mixed to produce 400 ml of the final reacting mixture, which is placed in eight Teflon vessels (50 ml each). These tubes are put inside the microwave oven, and the synthesis reaction was carried out using the same reaction conditions as reaction 1. Finally, the product was filtered and washed with distilled water.

X-ray diffraction analysis was performed to obtained the product, and again a single crystalline phase was obtained, and it was identified with the ICCD PDF #86-1199 (**Figure 1**(**b**)). The crystalline quality was improved in comparison to that showed by the HAp in reaction 1. Observation by SEM was also done to the sample, and an evident change in the morphology could be seen. This time the HAp possessed the morphology of nanowires as shown in **Figure 1**(**b**).

Reaction 3 was done with a GA concentration of 0.9 CNC. The reaction conditions were the same as those in reaction 1. The synthesized product from this reaction also showed a single crystalline phase identified as PDF #86-1199, but this time the Bragg reflection corresponding to the planes (300) at 32.92° in 2θ showed an increased intensity according to that reported in the mentioned PDF (as shown in **Figure 1**(**c**)). This is an evidence of a preferential crystalline orientation. The crystalline quality was increased again. The observation done by means of SEM revealed the morphology of well-defined fibers (**Figure 1c**).

For reaction 4, once again, the GA concentration was increased to 2 CNC, and the results were rather significant. This time, the X-ray diffraction showed a notorious increment in the intensity of the Bragg reflection produced by the (300) at 32.92°. This intensity is even higher than that for the (211), which is the most intense Bragg reflection according to the ICDD PDF #86-1199 (**Figure 1**(**d**)).

Similarly, the Bragg reflections produced by the planes (100) and (200) also experienced an increment in their intensities but in the diffractogram depicted in **Figure 1**(**d**), it is also possible to see a reduction in the intensity of the signal at 26°, which is generated by the (002). These changes in the Bragg reflection intensities are indicative of a remarkable preferential crystalline orientation on the HAp crystal structure.

Observations made using the SEM allowed seeing the morphology in the HAp obtained in this reaction, and the microscope revealed the morphology of microfibers as shown in **Figure 1**(**d**). This time, the microfibers were larger and thicker than those obtained in reaction 3, and they showed a quite smooth surface, their facets and ends looked well-defined, which is an evidence of a good-quality crystal growth. Additionally, all fibers showed a hexagonal cross section, their average diameter was 4.67 μm, and their lengths were in the range of hundreds of micrometers.

This morphology of fiber with a hexagonal profile allowed understanding the XRD results is depicted in **Figure 1**(**d**). During sample preparation for XRD experiments, most of the fibers laid on the sample holder and they acquired an arrangement in such way that their hexagonal profiles were perpendicular to the sample holder surface and their lengths were parallel to it (see **Figure 2**). As a result, most of the reciprocal vectors a* of the HAp crystal structure laid orthogonal to the surface and the possibilities to satisfy the diffraction conditions were increased. In contrast, a large number of reciprocal vectors c* laid parallel to the sample surface, and the probability to satisfy the Bragg conditions decreased considerably. Therefore, this fibers distribution made possible that the Bragg reflections produced by those planes orthogonal to the "a" axis in the HAp crystal structure (or "b" axis) had bigger intensity; In

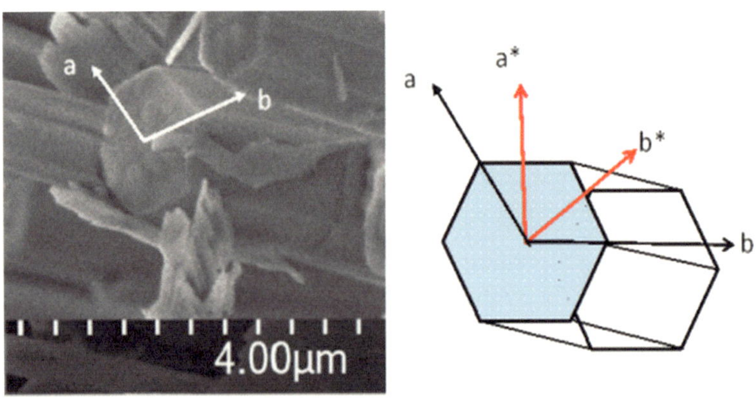

Figure 2. SEM micrograph showing the hexagonal profile in the HAp microfiber (left) and the schematic representation of the vectors "a" and "b" in the direct space, and the vectors "a*" and "b*" in the reciprocal space (right).

contrast, the reflection generated by the planes perpendicular to the "c" axis decreased their intensity significantly in the X-ray diffraction experiments. Consequently, in the diffractogram shown in **Figure 1(d)**, the contributions of the (300), (200) and (100) planes possess higher intensities and that corresponding to the (002) plane is nearly null.

In a closer observation on the hexagonal cross section of the fiber in **Figure 1(d)**, it was possible to see the microfiber was formed of small fibers closely packed within its body. These small fibers possessed pointed-end structures as can be seen in **Figure 1(d)** and the right micrograph in **Figure 3**. Besides, some of the microfibers also had a pointed-end structure as it was depicted in the left image in **Figure 3**.

In the SEM micrograph on the right side of **Figure 3**, it is also possible to see a hole in the center of the microfiber profile, this empty space could be produced when some loose nanofibers slid out of the microfiber body, and then those free nanofibers were finally situated near of the microfiber. It is possible that the lone nanofiber indicated by the white arrow could have undergone a slide as described before.

Observation of nanofibers using high-resolution transmission electron microscopy (HRTEM) revealed that they had an average diameter of 97.5 nm, and a polycrystalline microstructure as can be seen in **Figure 4(b)** and **(c)**. The average crystallite size was of 9.76 nm, which was calculated using the Digital Micrograph software.

Using that software, it was possible to obtain fast Fourier transforms (FFT) from the high-resolution images to analyze the crystal structure of the HAp in the nanofibers. The interplanar distances were also determined to identify the crystalline planes that contributed to form the images. In **Figure 5**, a bright field image of a nanofiber is depicted. In the upper right inset of **Figure 5**, a high-resolution micrograph of the nanofiber tip was showed. In this micrograph, it was possible to observe an arrangement of lines and dots, which were produced by the HAp crystal structure. The distances between lines were determined and the value of 0.344 nm

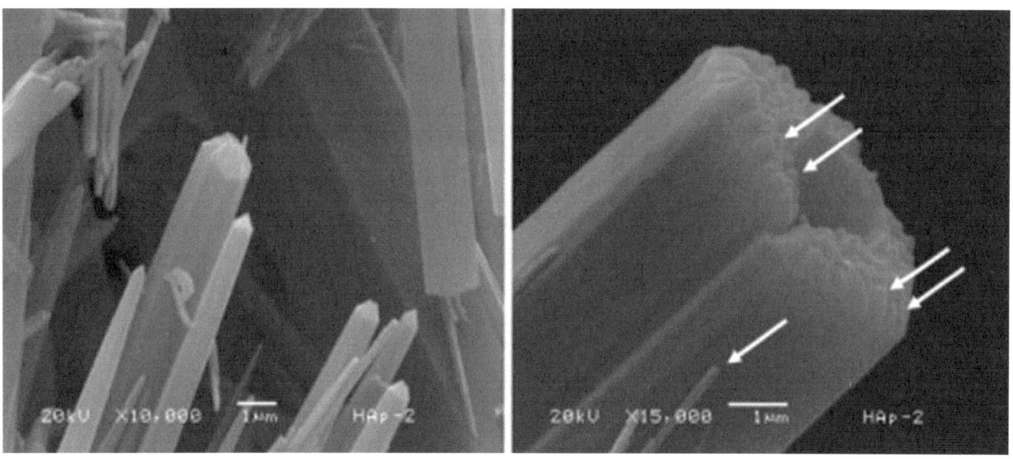

Figure 3. SEM micrographs of HAp microfibers showing the pointed-end (left) and how those nanofibers joint to form the microfiber structure (right), the white arrows indicate some nanofibers.

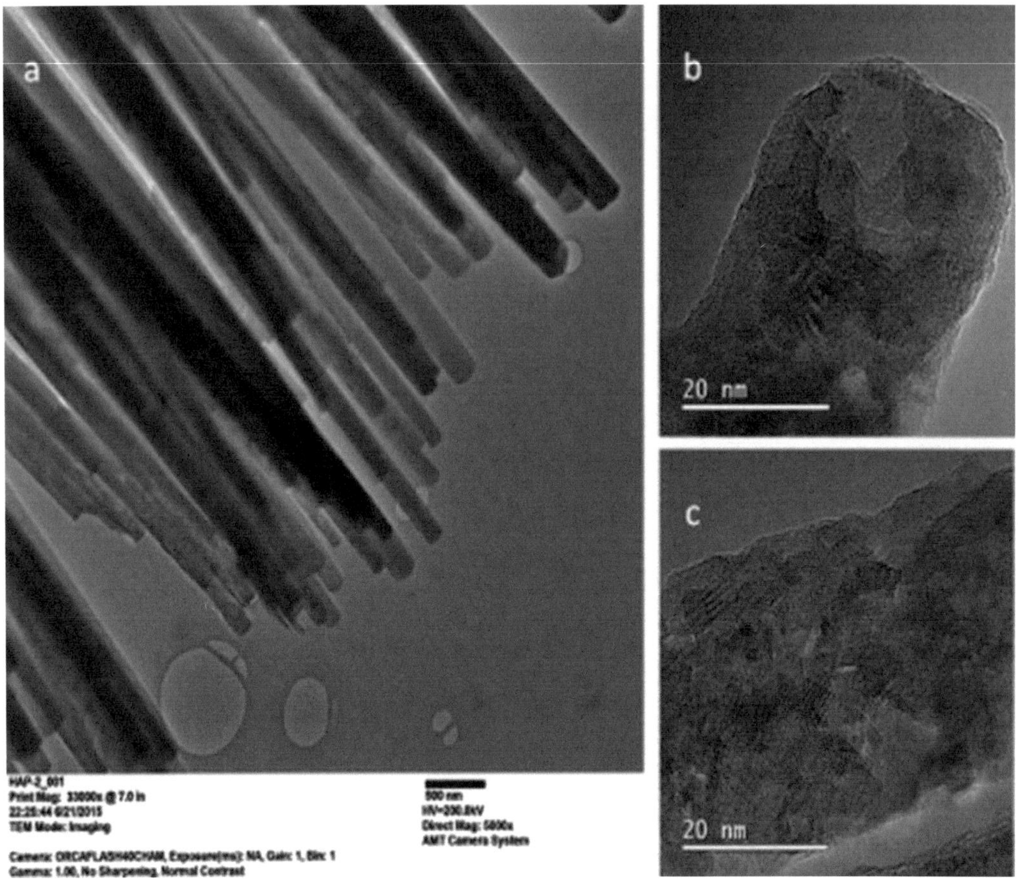

Figure 4. HRTEM images of nanofibers depicting their diameter size (a), and the polycrystallinity in their structure (b) and (c).

corresponds to the interplanar distance of (002) planes, and the distances between dots were of 0.295 nm, which were related to the (211) planes. A FFT was produced from the high-resolution micrograph and the dot pattern displayed in the bottom right inset was obtained.

The planes (002) generated the most intense dots in the pattern and this intensity is indicative of their abundance, which was also corroborated by the high-resolution micrograph in the upper inset of **Figure 5**. Other dots were produced by the planes (211). These crystalline planes are the most abundant according to the PDF 86-1199, but they are not numerous in the HAp obtained in this experiment. All dots in the FFT were laid on parallel lines and their arrangement is indicative of a high crystallinity. All (002) planes were stacked along the length of the nanofiber, and this stacking line was parallel to the "c" axis of the HAp crystal structure, this is indicative that the nanostructure was grown in the [002] direction as it was indicated in the upper right inset in **Figure 5**. These results were very significant because the preferred crystalline orientation found in the X-ray diffraction analysis, in the [300] direction, along with the SEM and HRTEM results indicates that the HAp fibers grow in [002] direction.

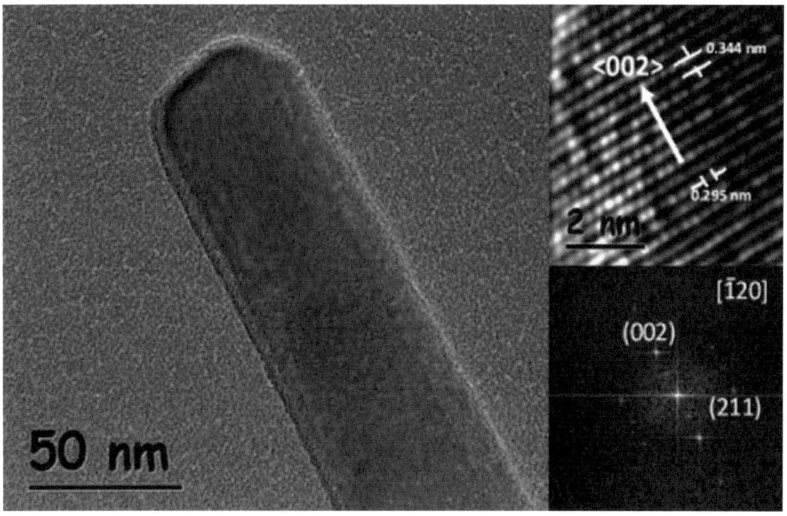

Figure 5. HRTEM image of a nanofiber showing its morphology. In the upper right inset, a high-resolution image of the fiber tip is displayed, the interplanar distances were calculated, and in the bottom right inset, a FFT was generated form the upper image and all dots were indexed.

These morphological and structural features of HAp fibers made them suitable to be used as the main material to elaborate porous ceramics, which must have porosity in the macro scale to allow the flow of nutrients and cells and consequently, a bone regeneration process could be possible.

Finally, once again the glutamic acid concentration increased in the reaction mixture and reaction 5 was performed to obtain another HAp structure. This time the concentration of glutamic acid was 250% related to the calcium nitrate content. The X-ray analysis of the product obtained in this reaction showed a small decrement on the intensity of the Bragg reflection corresponding to the (300) planes, and an increase in the intensity of the Bragg reflection at 26° related to the (002) planes, this is observed in the diffractogram in **Figure 6(a)**. The latest is indicative that the preferential crystalline orientation is lesser in this product in comparison with the microfibers described before.

The HAp obtained in this reaction was observed by means of the SEM and the morphology of nanoplates was found. These nanoplates had different shape in the plane surface, and they had an average thickness of 62.3 nm. This morphology can be seen in **Figure 6(b)**.

Similar to the case of the HAp microfibers, the nanoplates were formed by several small tickertapes or nanoribbons, which were joined to build the nanoplate surface (see **Figure 6(c)**). Besides, the thickness of these nanoplates were also formed by the stacking of those small nanoribbons in **Figure 6(c)** is possible to see some layers on the nanoplate surface and they were produce by the superposition of those nanoribbons. May be the whole nanoplate volume was formed by the union of several nanoribbons.

Possibly, during the crystal growth mechanism, the nanofibers or nanoribbons were formed, depending on the GA concentration in the reacting mixture, and their growing continued until

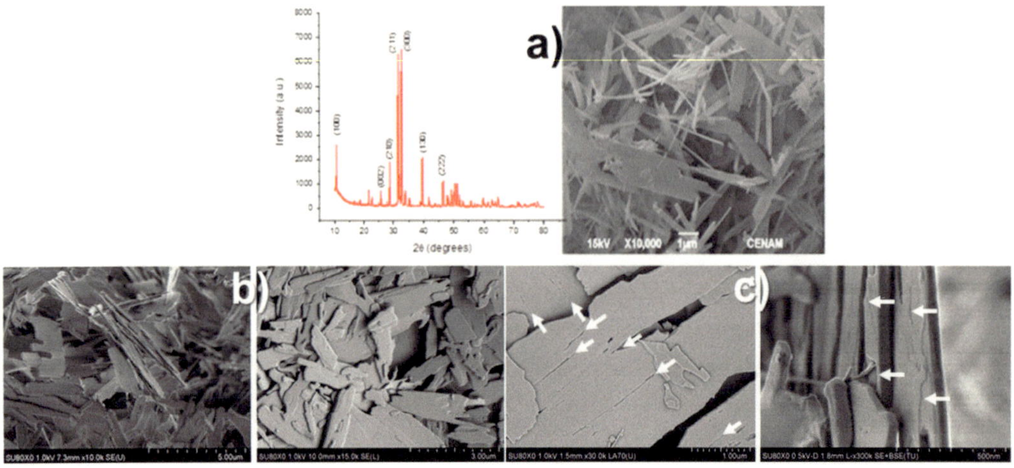

Figure 6. X-ray diffractogram of the HAp synthesized by the fifth chemical reaction and the morphology of nanoplates observed (a). SEM micrographs of nanoplates showing different shapes and sizes on the flat surfaces (b) and (c). The white arrows indicate the union between nanoribbons to form the flat surface, and the stacking of those nanoribbons to form the nanoplate thickness.

these nanostructures were linked to form the microfiber or nanoplate volume, respectively. The energy available within the reacting solution, especially in the regions close to the outer surface of the nanostructure, could be fused with the nanostructures to form a smooth surface. When the microfiber or nanoplate surface was degraded or corroded, it revealed that it was formed by the union of smaller pieces, corresponding to the nanofiber or nanoribbons.

5. Elaboration of HAp ceramics with interconnected porosity

Porosity plays a decisive role in the behavior of biomaterials. It is necessary to have sufficient pore size and interconnected porosity to promote osteoblasts to grow into a ceramic device [45, 46] and to allow cell penetration, internal tissue growth, vascular incursion and nutrient supply.

In order to manufacture HAp objects with different (and desired) shapes and sizes, which possess controlled and interconnected porosity, enough to permit cellular mobility and tissue ingrowth, it has used a successful process, named as modified gelcasting process (MGCP), previously reported by our group [47–50]. The key process is the proper use of a monomer-polymer blend, which can function as plasticizers, binders and dispersants and holds HAp powders or fibers together. During the polymerization, the slurry can be poured into silicone molds and a green body is obtained. PMMA micro balls (10–40 μm in diameter) were used as sacrificial porogens and oxalic acid was used to produce CO_2 bubbles during the polymerization to form the microporosity. Once the ceramic pieces are demolded, they are subjected to a thermal treatment in order to remove the polymers and promote a sintering of the HAp particles or fibers to obtain a ceramic with interconnected and controlled porosity. **Figure 7** shows HAp ceramic pieces with different shapes and sizes obtained through the MGCP.

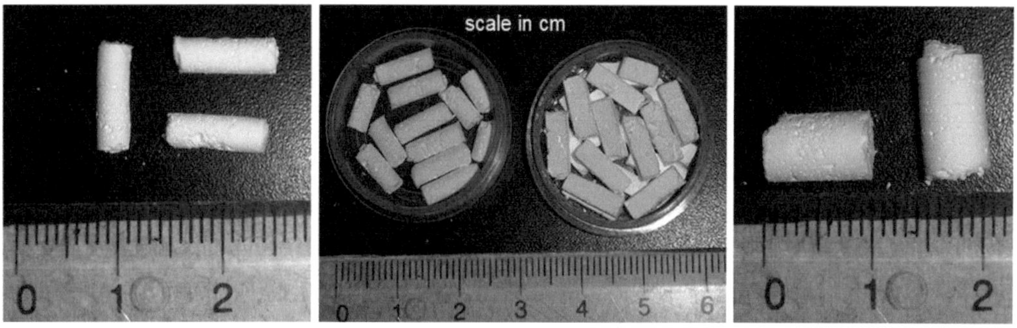

Figure 7. Hydroxyapatite objects molded in different sizes and shapes by the modified gelcasting process (scale in cm).

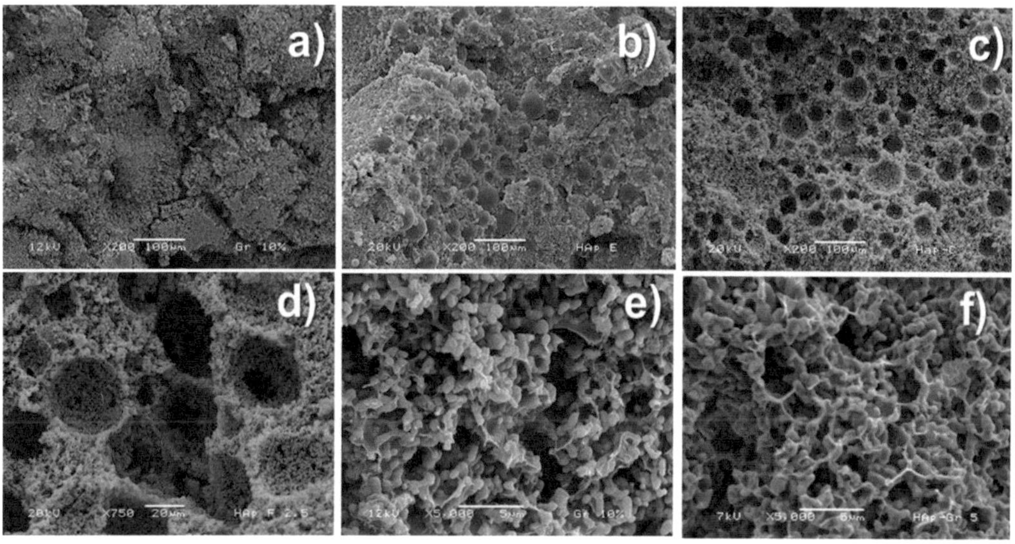

Figure 8. SEM micrographs of samples after gelcasting without PMMA micro balls (a) and with PMMA micro balls (b). SEM micrographs of HAp scaffold after polymer removal showing interconnected micro and macro porosity (c) and (d), and typical SEM micrographs of the HAp-based biomimetic organic-inorganic, composite material. The organic phase wet and links the HAp inorganic structure (e) and (f).

Figure 8(**a**) and (**b**) shows SEM micrographs of two demolded samples after the gelcasting process, in which polymers have not been eliminated. The difference between these two samples consists in that in the one of **Figure 8**(**a**) there is no PMMA micro balls, meanwhile in the sample of **Figure 8**(**b**) they can be clearly observed occupying spherical spaces that will generate the larger (or macro) porosity in the final HAp ceramic.

After the thermal treatment, polymers were removed and HAp fibers (or particles) are sintered to form a ceramic with interconnected porosity in micro (1–5 μm) and macro (20–40 μm) scales. This is relevant for the application as bone replacement because the osteocyte cells have a size between 100 and 500 nm [51], while osteoclast cells have a size of about 10 μm [52], which means that there is sufficient space to move through the porous structure. **Figure 8**(**c**) and (**d**) show SEM micrographs in two different magnifications in which the interconnected micro and macro porosity is observed.

6. Obtaining the HAp-based organic-inorganic composite material

Due to the fact that the combination of polymers and hydroxyapatite to fabricate bone substitutes is a natural strategy, in order to elaborate a HAp-based organic-inorganic composite material, a water solution of gelatin (or collagen) was used as organic phase. HAp ceramics with interconnected and controlled porosity, obtained as described earlier, were used as inorganic phase and the final composite material was obtained by following the steps described in **Figure 9**.

Different water solutions of gelatin with 1.5, 2.5, 5, 6.75 and 7.5 wt%, were dripped to ceramic scaffolds to generate an organic-inorganic composite material. SEM analyzed the incorporation of gelatin (collagen) to the matrix and a typical example is shown in **Figure 8**(**e**) and (**f**). It is interesting to notice that the organic phase wet and links the HAp inorganic structure to form the organic-inorganic composite material in a similar way as in the natural bone tissue.

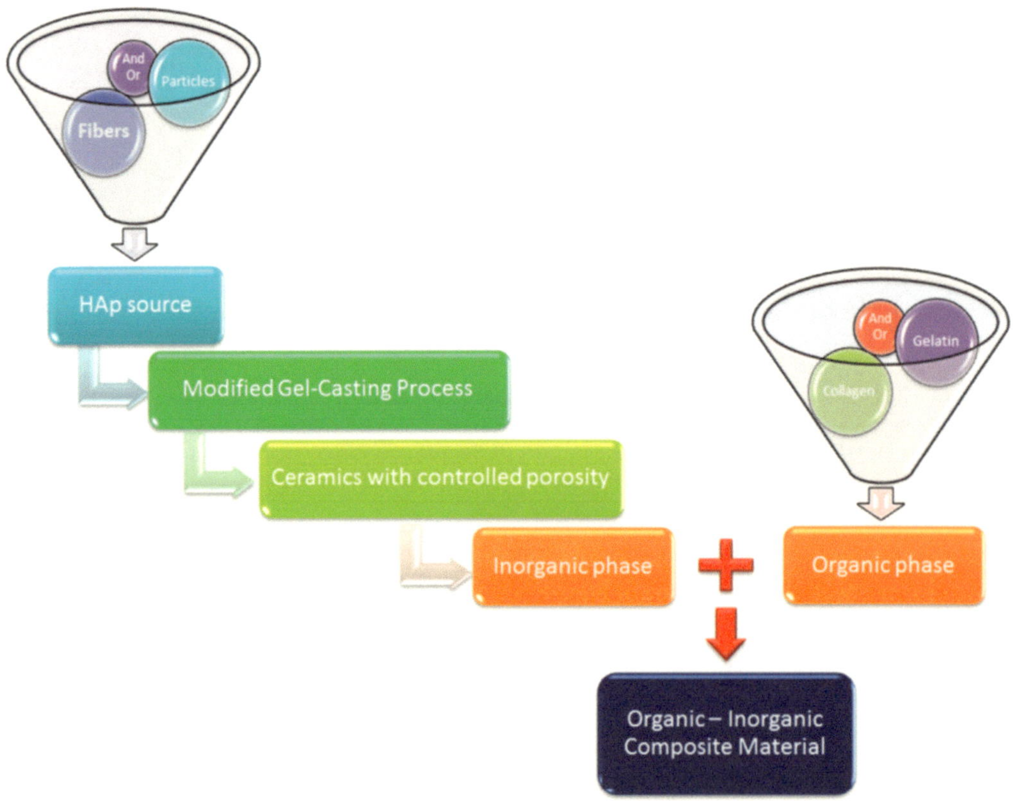

Figure 9. General procedure to obtain HAp-based organic-inorganic composite material.

It has been established that a complex between calcium ions (Ca^{2+}) in HAp and —COO— groups in the amino acids in the organic phase is formed, which means that a chemical interaction between the organic and inorganic phases occurs to form the composite material [49, 50]. These interactions will play an important role in improving the mechanical properties of the final material.

Mechanical properties were evaluated under compression. **Figure 10(a)** shows a typical mechanical behavior of composite samples which corresponds to a typical non-linear elastic behavior showed by cellular materials and which agrees completely with the observations of SEM images as shown in **Figure 8(c)** and **(d)**. The mechanical behavior shows two stages; in each of them, a linear behavior, corresponding to an elastic deformation is observed, after which a change in the slope of the curve is observed. At that point, the cells are deformed plastically until they reach the maximum compressive strength and begin to collapse, dissipating energy by breaking the bonds between the Ca^{2+} ions of the HAp and the —COO— groups of the organic phase mentioned above. Then, a densification of the material follows until the second stage is reached and the behavior is repeated until reaching a second maximum compressive strength (with a lower value). The highest of these maximum is reached due to the macroporosity while the following are reached due to the microporosity. Similar behavior has been described in different materials and studied by Maskery et al. [53].

The maximum compressive strengths were determined as a function of the gelatin concentration, as can be observed in **Figure 10(b)**, and it is interesting to notice that there is a dramatic increase in the maximum compressive strength for the composite materials compared with the sample in which no organic phase was added (marked as WOP). Moreover, the maximum compressive strength increases as the amount of gelatin in the composite material increases, due to a synergic behavior between the inorganic and organic phases, to reach a value of 18 MPa for the sample with 7.5 wt% of gelatin. The aforementioned Ca—O bonds can absorb energy as sacrificial bonds, resulting in a synergic contribution to improve the mechanical properties of this composite material as well as due to the morphology of the HAp fibers. It is clear that for a hexagonal-shaped crystal, the direction in which it can withstand greater mechanical stresses is precisely [001] parallel to the axis of the hexagon. Consequently, the

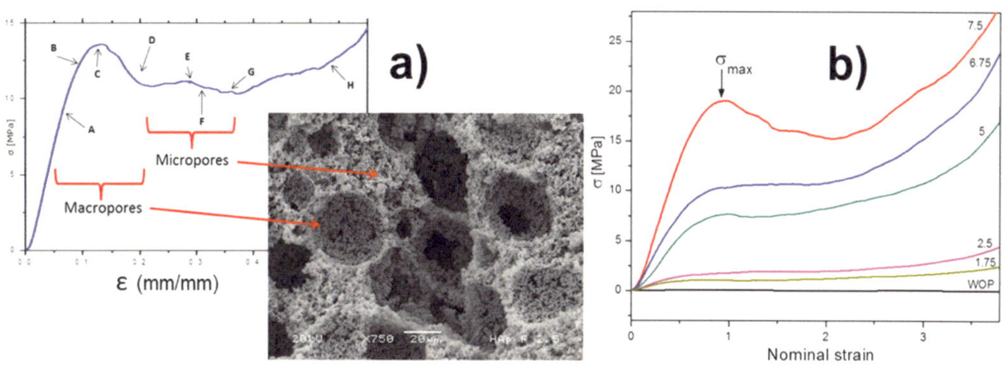

Figure 10. Mechanical compressive behavior of HAp-based biomimetic composite material (a). Stress-strain curves obtained in compressive tests for samples with different concentrations (wt%) of gelatin (b).

morphology of the HAp fibers obtained by the MAHM (grown with preferential orientation in the direction [001]), positively influences the improvement of the mechanical properties of the final HAp-based organic-inorganic composite material.

This value of 18 MPa, is 3.2 times higher than that reported for cancellous bone, and in the same order of magnitude for the cancellous-cortical bone tested under the same conditions [54], which implies that this biomimetic organic-inorganic composite material could be suitable for bone replacement.

7. Conclusions

After all experiments to synthesize different HAp nanostructures, it is possible to assure that the GA in the reacting mixture had a relevant impact on the final morphology of HAp clusters. This amino acid guided the HAp crystal growth and also leads the arrangement of small structures to form the final morphology, which was dependent on the GA concentration. In addition, this amino acid also allowed to obtain a preferential crystalline orientation in the [001] direction of the HAp crystal structure in most of the nanostructures obtained. The use of the MAHM made the synthesis of HAp with a high crystalline quality and purity easier.

Nanofiber was the most relevant morphology obtained in the synthesis reactions, and the arrangement of those to form microfibers with a hexagonal profile was unique. This morphology was selected to be used in the production of HAp porous ceramics due to its similarity with that of HAp morphology present in the human bone tissue.

The porous ceramic was successfully obtained through the MGP using the HAp microfibers. The micro and macro porosity were controlled in the ceramic body by the use of different polymers and chemical substances during the production process, and the pore size distribution was suitable to permit the flow of nutrients and cells.

The addition of protein to the porous ceramic allowed producing organic-inorganic composite materials in a biomimetic way. The strong chemical interaction between both phases leads to synergic mechanical properties in the composite materials.

Mechanical tests showed that the composite materials had a similar mechanical strength to that observed in the trabecular bone. These results suggest that this composite material can be an excellent option to be used as bone replacement or implant materials. In addition, the characterization of those composite materials proved that they also had an interconnected porosity in the micro and macro scales, which was proper to permit the natural regeneration processes of the human bone tissue.

Acknowledgements

Authors acknowledge the financial support of CONACYT scholarships and Fondo Salud 2013-01-201329. Authors also acknowledge Alicia del Real López and Carmen Peza Ledesma for

http://dx.doi.org/10.5772/intechopen.71850

technical assistance in SEM analysis and Gilberto Mondragón Galicia and Jesús Arenas Alatorre for technical assistance in HRTEM studies.

Author details

Eric M. Rivera-Muñoz[1]*, Rodrigo Velázquez-Castillo[2], Susana Alonso-Sierra[2], J. Rafael Alanís-Gómez[2], Beatriz Millán-Malo[1], Lauro Bucio-Galindo[3], Rafael Huirache-Acuña[4], Alejandro Manzano-Ramírez[5], Rufino Nava[2] and Miguel Apátiga-Castro[1]

*Address all correspondence to: emrivera@fata.unam.mx

1 Center for Applied Physics and Advanced Technology, National Autonomous University of Mexico, Querétaro, Qro., México

2 Division of Research and Postgraduate, Faculty of Engineering, Autonomous University of Querétaro, Querétaro, Qro., México

3 Institute of Physics, National Autonomous University of Mexico, México, D.F., México

4 Chemical Engineering Faculty, Michoacan University of San Nicolás de Hidalgo, Ciudad Universitaria, Morelia, México

5 CINVESTAV-Querétaro, Libramiento Norponiente # 2000, Fraccionamiento Real de Juriquilla, Querétaro, México

References

[1] Campa J, Ulloa S, Bucio L, Belio IA, Velazquez R, Rivera E. Biomateriales: Fundamentos, técnicas y aplicaciones. 1st ed. Mexico: Universidad de Guadalajara; 2007. 127 p

[2] Avashnee C, Ilse W, Marei MK, Yasser EK, Moussa Rania M. Synthesis, properties, and applications of Please provide location name for the publisher in Ref. [2].hydroxyapatite. In: Gshalev Valeri S, Demirchan Aleksandra C, editors. Hydroxyapatite: Synthesis, Properties and Applications. Nova Sciences Publishers; 2012. pp. 91-132

[3] Silva CC, Pinheiro AG, Figueiró SD, Góes JC, Sasaki JM, Miranda MAR, Sombra ASB. Piezoelectric properties of collagen-nanocrystalline hydroxyapatite composites. Journal of Materials Science. 2002;**37**:2061-2070

[4] Bowen CR, Gitting J, Turner IG, Baxter F, Chaudhuri JB. Dielectric and piezoelectric properties of hydroxyapatite-$BaTiO_3$ composites. Applied Physic Letters. 2006;**89**(13):132906

[5] Bowen CR, Raman KVS, Topolov VY. Piezoelectric composites based on hydroxyapatie/barium titanate. Advances in Science and Technology. 2008;**54**:1-6

[6] Silva CC, Thomazini D, Pinheiro AG, Aranha N, Figueiro SD, Góes JC, Sombra ASB. Collagen-hydroxyapatite films: Piezoelectric properties. Materials Science & Engineering B. 2001;**86**(3):210-218

[7] Jianqing F, Huipin Y, Xingdong Z. Promotion of osteogenesis by a piezoelectric biological ceramic. Biomaterials. 1997;**18**(23):1531-1534

[8] Ding Y, Liu J, Wang H, Shen G, Yu RA. Piezoelectric immunosensor for the detection of a-fetoprotein using an interface of gold/hydroxyapatite hybrid nanomaterial. Biomaterials. 2007;**28**(12):21147-22154

[9] Damjanovic D. Ferroelectric, dielectric and piezoelectric properties of ferroelectric thin films and ceramics. Reports on Progress in Physics. 1998;**61**(9):1267-1272

[10] Tsuchida T, Yoshioka T, Sakuma S, Takeguchi T, Ueda W. Synthesis of biogasoline from ethanol over hydroxyapatite catalyst. Industrial & Engineering Chemistry Research. 2008;**47**(5):1443-1452

[11] Xu J, White T, Li P, He C, Han YF. Hydroxyapatite foam as a catalyst for formaldehyde combustion at room temperature. Journal of the American Chemical Society. 2010;**132**(38): 13172-13173

[12] Tsuchida T, Kubo J, Yoshioka T, Sakuma S, Takegushi T, Ueda W. Reaction of ethanol over hydroxyapatite affected by Ca/P ratio of catalyst. Journal of Catalysis. 2008;**259**(2): 183-189

[13] Sebti S, Tahir R, Nazih R, Boulaajaj S. Comparison of different lewis acid supported on hydroxyapatite as new catalysts of Friedel-craft alkylation. Applied Catalysis A: General. 2001;**218**(1–2):25-30

[14] Ogo S, Onda A, Yanagisawa K. Selective synthesis of 1-butanol from etanol over strontium phosphate hydroxyapatite catalysts. Applied Catalysis A: General. 2011;**402**(1–2): 188-195

[15] Chackraborty R, Bepar Si, Banerjee A. Application of calcined waste fish (Labeo rohita) scale as low-cost heterogeneous catalyst for biodiesel synthesis. Bioresource Technology. 2011;**102**(3):3610-3618

[16] Ngamcharussrivichai C, Nunthasanti P, Tanachai S, Bunyakiat K. Biodiesel production through transesterification over natural calciums. Fuel Processing Technology. 2010;**91** (11):1409-1415

[17] Jiang S-D, Yao Q-Z, Zhou G-T, Sheng-Quan F. Fabrication of hydroxyapatite hierarchical hollow microspheres and potential application in water treatment. Journal of Physical Chemistry. 2012;**116**(7):4484-4492

[18] Ozawa M, Hattori M, Satake K. Waste management and application of fish bone hydroxyapatite for waste water treatment. In: Proceedings of the International Symposium on Ecotopia Science 2007; November 23–25, 2007; Nagoya, Japan. Nagoya University; 2007. p. 957-958

[19] Wang F, Guo Y, Wang H, Yang L, Wang K, Ma X, Yao W, Zhang H. Facile preparation of hydroxyapatite with a three dimensional architecture and potential application in water treatment. CrystEngComm. 2011;**13**:5634-5637

[20] Boissona J, Ruttensb A, Mencha M, Vangronsveldb J. Evaluation of hydroxyapatite as a metal immobilizing soil additive for the remediation of polluted soils. Part 1. Influence of hydroxyapatite on metal exchangeability in soil, plant growth and plant metal accumulation. Environmental Pollution. 1999;**104**(2):225-233

[21] Seaman JC, Arey JS, Bertsch PM. Immobilization of nickel and other metals in contaminated sediments by hydroxyapatite addition. Journal of Environmental Quality. 2000;**30**(2): 460-469

[22] Jang SH, Jeong YG, Min BG, Lyoo WS, Lee SC. Preparation and lead ion removal property of hydroxyapatite/polyacrylamide composite hydrogels. Journal of Hazardous Materials. 2008;**159**(2–3):294-299

[23] Liu DM, Troczynski T, Tseng WJ. Water-based sol-gel synthesis of hydroxyapatite: Process development. Biomaterials. 2001;**22**(13):1721-1730

[24] Padmanabhan SK, Balakrishnan A, Chu M, Lee YJ, Kim TN, Cho SJ. Sol-gel symthesis and characterization of hydroxyapatite nanorod. Particuology. 2009;**7**(6):466-470

[25] Liu DM, Yang Q, Troczynski T, Tseng WJ. Structural evolution of sol gel derived hydroxyapatite. Biomaterials. 2002;**23**(7):1679-1687

[26] Baba AA, Oduwole IT, Salami FO, Adekola FA, Adeboye SE. Symthesis of hydroxyapatite from waste egg-shell by precipitation method. Ife Journal of Science. 2013;**15**(3):435-443

[27] Rivera E, Araiza M, Brostow W, Castaño V, Díaz JR, Hernández R, Rodríguez R. Synthesis of hydroxyapatite from eggshells. Materials Letters. 1999;**41**(3):128-135

[28] Pramanik S, Agarwal AK, Rai KN, Garg A. Development of high strength hydroxyapatite by solid-state-sintering process. Ceramics International. 2007;**33**(3):419-426

[29] Guo X, Yan H, Zhao S, Li Z, Li Y, Liang X. Effect of calcining temperature on particle size of hydroxyapatite synthesized by solid-state reaction at room temperature. Advanced Powder Technology. 2013;**24**(6):1034-1038

[30] Yan L, Li Y, Deng ZX, Zhuang J, Sun X. Surfactant assisted hydrothermal synthesis of hydroxyapatite nanorods. International Journal of Inorganic Materials. 2001;**3**(7):633-637

[31] Wang Y, Chen J, Wei k, Zhang S, Wang X. Surfactant-assisted synthesis of hydroxyapatite particles. Materials Letters. 2006;**60**(27):3227-3231

[32] Goudarzi A., Solati-Hashjin M., Moztarzadeh F. Surfactant assisted synthesis of hydroxyapatite nano-rods by aqueous precipitation and hydrothermal post-treatment. In: Jürgen G. Heinrich, editor. Proceedings of the 10th ECerS Conference; June 17–21, 2007; Berlin, Germany. Baden-Baden Göller; 2008. p. 964-968

[33] Zhang HG, Zhu Q, Wang Y. Morphologically controlled synthesis of hydroxyapatite with partial substitution of fluorine. Chemistry of Materials. 2005;**17**:5824-5830

[34] Liu HS, Chin TS, Lai LS, Chiu SY, Chung KH, Chang CS, Lui MT. Hydroxyapatite synthesized by a simplified hydrothermal method. Ceramamics International. 1997;**23**:19-25

[35] Liu J, Ye X, Wang H, Zhu M, Wang B, Yan H. The influence of pH and temperature on the morphology of hydroxyapatite synthesized by hydrothermal method. Ceramics International. 2003;**29**:629-633

[36] Hui P, Meena SL, Singh G, Agarawal RD, Prakash S. Synthesis of hydroxyapatite bioceramic powder by hydrothermal method. Journal of Mineral & Materials Characterization and Engineering. 2010;**9**(8):683-692

[37] Earl JS, Wood DJ, Milne SJ. Hydrothermal synthesis of hydroxyapatite. Journal of Physics: Conference Series. 2006;**26**:268-271

[38] Wang KW, Zhu YJ, Chen F, Cheng GF, Huang YH. Microwave-assisted synthesis of hydroxyapatite hollow microspheres in aqueous solution. Materials Letters. 2011;**65**(15–16): 2361-2363

[39] Qi C, Tang QL, Zhu YJ, Zhao XY, Chen F. Microwave-assisted hydrothermal rapid synthesis of hydroxyapatite nanowires using adenosine 5-triphosphate disodium salt as phosphorous source. Materials Letters. 2012;**85**:71-73

[40] Cao JM, Feng J, Deng SG, Chang X, Wang J, Liu JS, Lu P, Lu HX, Zheng MB, Zhang F. Microwave assisted solid-state synthesis of hydroxyapatite nanorods at room temperature. Journal of Materials Science. 2005;**40**(23):6311-6311

[41] Cabrera JL, Velázquez R, Rivera-Muñoz E. Synthesis of hydroxyapatite nanostructures using microwave heating. Journal of Nanoscience and Nanotechnology. 2011;**11**(6):5555-5561

[42] Meejo S, Maneeprakorn W, Winotai P. Phase and thermal stability of nanocrystalline hydroxyapatite prepared via microwave heating. Thermochimica Acta. 2006;**447**:115-120

[43] Kalita SJ, Verma S. Nanocrystalline hydroxyapatite bioceramic using microwave radiation: Synthesis and characterization. Materials Science and Engineering C. 2010;**30**:295-303

[44] Han JK, Song HY, Saito F, Lee BT. Synthesis of high purity nano-sized hydroxyapatite powder by microwave-hydrothermal method. Materials Chemistry and Physics. 2006;**99** (2–3):235-239

[45] Joschek S, Nies B, Krotz R, Göpferich A. Chemical and physicochemical characterization of porous hydroxyapatite ceramics made of natural bone. Biomaterials. 2000;**21**:1645-1658

[46] He L-H, Standard O, Huang T, Latella B, Swain M. Mechanical behavior of porous hydroxyapatite. Acta Biomaterialia. 2008;**4**:577-586

[47] Rivera-Muñoz E, Díaz J, Rodríguez R, Brostow W, Castaño V. Hydroxyapatite spheres with controlled porosity for eye ball prosthesis: Processing and characterization. Journal of Materials Science Materials in Medicine. 2001;**12**:305-311

[48] Rivera-Muñoz E, Velazquez R, Rodriguez R. Improvement in mechanical properties of hydroxyapatite objects with controlled porosity made by modified gelcasting process. Materials Science Forum. 2003;**426-432**:4489-4494

[49] Rivera-Muñoz EM, Velázquez R, Muñoz-Álvarez P. Mechanical characterization of hydroxyapatite-based, organic-inorganic composites. Materials Science Forum. 2007;**539-543**:583-588

[50] Alonso-Sierra S, Velázquez-Castillo R, Millán-Malo B, Nava R, Bucio L, Manzano-Ramírez A, Cid-Luna H, Rivera-Muñoz EM. Interconnected porosity analysis by 3D X-ray microtomography and mechanical behavior of biomimetic organic-inorganic composite materials. Materials Science and Engineering C. 2017;**80**:45-56

[51] Su BL, Sánchez C, Yang X-Y. Hierarchically Structured Porous Materials: From Nanoscience to Catalysis, Separation, Optics, Energy, and Life Science. Wiley-VCH Berlag GmbH & Co: Weinheim, Germany; 2011. 678 p

[52] Gough J, Notingher I, Hench L. Osteoblast attachment and mineralized nodule formation on rough and smooth 45S5 bioactive glass monoliths. Journal of Biomedcal Materials Research Part A. 2004;**68A**:640-650

[53] Maskery I, Aboulkhair N, Aremu A, Tuk C, Ashcroft I, Wildman R, Hague R. A mechanical property evaluation of graded density Al-Si10-Mg lattice structures manufactured by selective laser melting. Materials Science and Engineering A. 2016;**670**:264-274

[54] Ravaglioli K. Bioceramics. Materials, Properties and Applications. Great Britain: Chapman & Hall; 1992. 422 p

Chapter 4

Synovectomy with ^{153}Samarium Hydroxyapatite in Haemophilic Arthropathy

José Ulisses Manzzini Calegaro

Additional information is available at the end of the chapter

http://dx.doi.org/10.5772/intechopen.71491

Abstract

There are two main types of haemophilia, classified according to deficiency: type A, caused by factor VIII deficiency; and type B, which is rarer and the result of a deficiency in factor IX. Haemarthroses account for 80% of bleeding in haemophilic patients, with half of these exhibiting deformities. Repeated joint effusion leads to a local inflammatory response, with the formation of hyperplasic and hypertrophic cells and subsequent buildup of haemosiderin. Fibroblasts proliferate and produce collagenases and proteinases that act on the synovium, cartilage and bone, with a decrease in the joint space. Another mechanism involved is the damage caused by direct action of red blood cells on the cartilaginous surface of the joint lining. 153-samarium was obtained in research reactor by neutron irradiation of $152Sm_2O_3$ (99.4%) in the nitrate form, 152Sm(n,p)153Sm, for 30–36 h. The labelling process was performed with 40 mg of hydroxyapatite, according to Barboza et al. Radiochemical purity, particle size, microbiological tests for sterility and pyrogen were the tests applied to obtain an useful material. The introduction of 153Sm-HA for the treatment of haemophilic arthropathy in large and medium joints was a safe, cost-effective, minimally invasive and effective procedure in controlling bleeding and pain.

Keywords: haemophilic arthropathy, radiosynovectomy, 153samarium-hydroxyapatite

1. Introduction

The treatment of joints using radioactive material began in the 1950s, more specifically in 1952, with chromic phosphate P32 [10]. It was initially aimed at joint involvement caused by rheumatoid arthritis and, to a lesser degree, pigmented villonodular synovitis, ankylosing spondylitis, collagenosis and psoriatic arthritis in the years that followed [24]. With the advent of longer follow-up studies, it has also benefitted rheumatic diseases and haemophilic arthropathy, which exhibit a similar sequence of events including repetitive intra-articular haemorrhages causing synovitis, joint pain, limited mobility and posterior muscular atrophy. Recurrent synovitis results in cartilage destruction, progressive loss of movement, joint deformities, bone damage and ultimately total ankylosis. The treatment procedure was originally denominated synovectomy (from the Greek 'ectomía' meaning 'to cut out') and later synoviorthesis (from the Greek word 'orthesis' meaning 'restoration') via radionuclides [8].

Different radioactive materials have been used to eradicate synovitis, some emitting only beta radiation and others beta and gamma radiation. Synovectomy in haemophilia using radioactive material began in 1971 [1]. Since then, a variety of materials have been used, including P32, colloidal 198Au, 186Re, 90Y, 165Dy, 166Ho, and 169Er. **Table 1** shows the characteristics of the materials used [1, 12, 16–18, 21, 28].

Irradiation occurs via the intra-articular retention of the radioactive material. However, it should be noted that the radioactive material is bound to larger particles, known as carriers, which undergo phagocytosis by the macrophages in the inflammatory process, favouring greater retention in the joint space. These macrophages migrate through the interstice of synovial cell layers, resulting in more homogeneous action by the ionising radiation. This behaviour was highlighted in autoradiographic studies [7], which more clearly indicate the location of samarium-153 particulate at different synovial tissue depths than other materials used, such as 186Re. This is also reported by Schneider et al. [23] (shown in **Figure 1**), in addition to direct irradiation by the intra-articular radionuclide. As such, average penetration is ascertained by the range of the β particle and maximum penetration by macrophage permeation into synovial cell layers.

Radioisotopes	Half-life (days)	Max. beta energy (MeV)	Gamma energy (KeV)	Penetration (mm)		Particle size (μm)	Leakage (%)
				Max.	Av.		
Colloidal 198Au	2.7	0.96	110	3.6	1.2	0.02–0.04	20–35
P32 (chromic phosphate)	14.0	1.7	–	7.9	2.6	0.05–0.1	2–4
186Re (sulphide colloid)	3.75	1.07	140	3.6	1.2	0.05–0.1	2
Colloidal 90Y	2.7	2.2	–	10.8	3.8	1.5–3.5	3
166Ho (FHMA)	1.2	1.85	81	8.7	2.2	1.82–12	1
165Dy (FHMA)	0.095	1.3	95	5.6	1.4	0.8–12	1
169Er (citrate)	9.4	0.34	–	1.0	0.3	0.1–10	1
153Sm (HA)	1.95	0.80	100	3.1	0.8	1–10	0.1

FHMA = ferric hydroxide macroaggregates.

Table 1. Characteristics of the radioisotopes used in synovectomy.

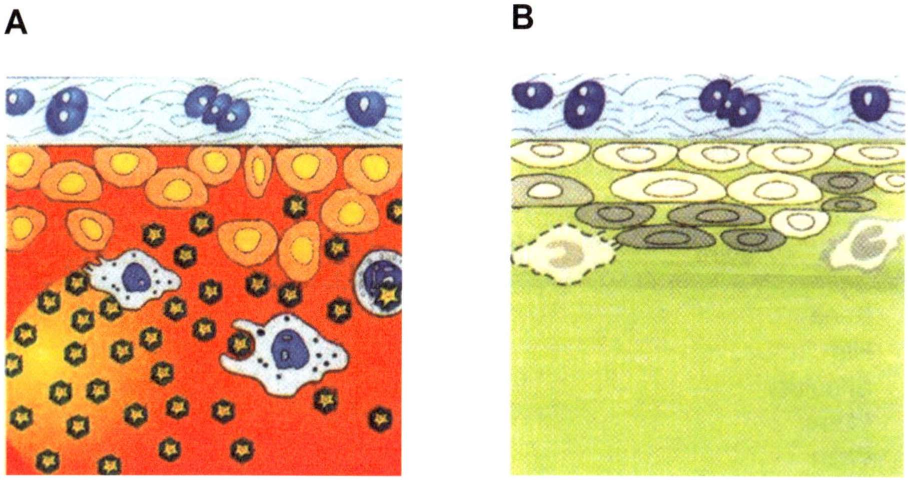

Figure 1. (A) β-Emitting colloidal particles (yellow stars) phagocytised by inflamed hypertrophic synovial lining with proliferating synoviocytes (pink). The cartilage layer remains unaffected. (B) Subsequent cell damage and sclerosis of synovial membrane.

Another noteworthy aspect is that radionuclide leakage from the joint is inversely proportional to particle size. This is clearly evident in the last two columns of **Table 1**. The presence of gamma radiation allows synovectomy to be monitored over several days via scintigraphy.

2. Haemophilic arthropathy

Haemophilia is a congenital bleeding disorder linked to the X chromosome of the human genome, with the two most common types being haemophilia A, a lack of blood clotting factor VIII, and haemophilia B, caused by missing of defective factor IX. Joint bleeding associated with muscle bleeding represents 90% of bleeding episodes in haemophilia patients, while haemarthrosis alone accounts for 70–80% of these episodes. In 80% of cases, haemarthrosis occurs in the knees, elbows and ankles [22], producing inflammatory changes in the synovial membrane. Recurrence of this inflammation over time triggers a chain of events that lead to joint ankylosis, including the direct damage of blood on the articular cartilage [15].

In cases of arthropathy mediated by reactive synovitis, synovectomy with radioactive material is an alternative to intra-articular injection of glucocorticoids and other chemical agents (osmic acid, collagenase, rifocin and thiotepa) or surgery. Furthermore, radiosynovectomy, introduced by Ahlberg in the 1970s, has been proposed as a first-line treatment option for haemophilic arthropathy [22, 23].

3. 153Samarium-hydroxyapatite

153-samarium was obtained in an IEA-R1 research reactor (IPEN-CNEN, São Paulo, Brazil) by neutron irradiation of $152Sm_2O_3$ (99.4%) in the nitrate form, 152Sm(n,p)153Sm, for 30–36 h. The labelling process was performed with 40 mg of hydroxyapatite, using appropriately sized particles (20 μm), according to Barboza et al. [2]. Percentage bound activity or labelling efficiency was determined by centrifugation, measuring the activity of the precipitate (153Sm-HA) and supernatant (153Sm-free) using a dose calibrator, and was always >90%. Radiochemical purity was higher than 98%, measured using Whatman 3MM paper chromatography (from GE, Milwaukee, WI, USA) in 0.9% saline, remaining stable for 24 h. Particle size was determined by laser scattering and filtration in a filter system of known sizes (1–15 μm), with a mean of 10 μm (range: 3–12 μm). Microbiological tests for sterility and pyrogen were always negative in all samples.

4. Patient selection

Participants were haemophilic patients with chronic synovitis, monitored at the Department of Haematology – Hospital de Apoio in the Federal District (DF), and the Orthopaedics and Nuclear Medicine Service of the Hospital de Base (DF), between 2002 and 2011.

Patients were assessed by clinical history and a physical examination, paying special attention to the compromised joint. Particular emphasis should be given to haematology tests in order to characterise the type and severity of haemophilia, as well as the absence of coagulation inhibition factors. Imaging (radiology) tests make it possible to determine the degree of arthropathy, with the Pettersson score being widely used. Three-phase whole-body scans are used to identify any other joints involved and more accurately characterise the synovial inflammatory process. Other procedures used for this purpose include ultrasound or magnetic resonance imaging (particularly for the knees).

Inclusion criteria were the following: chronic synovitis when occurs repetitive effusions (minimum: once a month), pain on joint palpation and absence of other joint disease, like the rheumatologic or orthopaedic nature. Exclusion criteria were ruptured Baker's cyst, major effusions, signs of acute synovitis or presence of an articular or periarticular infectious process.

5. Procedure

Synovectomy of the radioactive material was made by an orthopaedist with previous use of the deficient clotting factor, applying topical anaesthesia in accordance with the asepsis and antisepsis performances used for invasive intra-articular orthopaedic procedures. The use of ultrasound to guide the punctures was not necessary, as these were assured by the aspiration of synovial fluid before administration of the radionuclide (**Figure 2**). Fixed doses of 5 mCi (185 MBq) or 20 mCi (740 MBq) of 153Sm-hydroxyapatite were used, with only one injection

http://dx.doi.org/10.5772/intechopen.71491

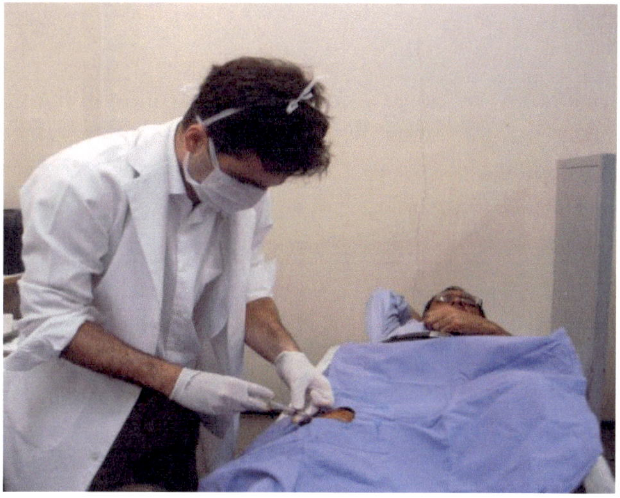

Figure 2. Injection of the knee.

administered per patient. The maximum volume of radioactive material was 0.5 mL; material adhered to the puncture site was washed away with saline, at fractions of 0.5 mL, without exceeding the final volume of 1.5–2.0 mL.

The reflux and homogeneity (or lack thereof) of the intra-articular material and its escape from the joint were monitored by a scintigraphic study in a gamma camera, with a wide field of vision detector and low-energy collimator. Imaging was made using a 128-pixel matrix and the spectrometer window was centred at 100 keV, using precocious, 1 and 2 h, and later times, 3–7 days, after 153Sm-HA injection. A summary of this protocol is shown in **Table 2**.

1	Use a coagulation factor before the procedure;
2	Local asepsis;
3	Use local anaesthetic;
4	Joint puncture using a 21G needle;
5	Aspirate synovial fluid;
6	Inject 5 mCi of 153Sm-HA into the intermediate joints and 20 mCi into the knees;
7	Wash the puncture site with a total of 2.0 mL of saline without using corticosteroids;
8	Compression bandaging using crepe bandage;
9	Monitor the material used in the puncture;
10	Functionally permissible joint mobility;
11	Immediate (1–2 hours after synovectomy) and later (3–7 days) scintigraphy image

Table 2. Protocol for synovectomy with 153Sm-hydroxyapatite.

6. Side effects

Reactive synovitis may occur in the days following application of the radiopharmaceutical agent, which was treated using conservative measure such as joint rest, local application of ice and a non-steroid anti-inflammatory agent. The cases observed in our study were mild and occurred in 4–8% of joints.

Radionecrosis may occur if the material leaks from the administration route. This complication did not occur in the cases we treated because the material was only administered after correct injection and the puncture site was washed with saline to prevent leakage.

Immobilisation of the affected limb will ultimately result in thrombosis; however, this was prevented by the protocol used.

There is no concern about possible carcinogenic effects of this procedure. Systemic irradiation can result from fluid leakage or the gamma component of 153Sm. Studies on chromosomal abnormalities in circulating lymphocytes related to samarium [19] have shown no definitive changes, but rather transient and reversible ones. It is important to note that this irradiation is smaller than in diagnosis by conventional bone scintigraphy or whole-body scanning with 67Ga. Considering the local effect at joint level, several studies with long follow-up times have shown no occurrence of tumours [14, 26, 27], indicating that this possibility has not yet been characterised.

7. Synovectomy

The first study was conducted to evaluate the efficiency of treatment with 153-samarium hydroxyapatite (153-Sm-HA) in haemophilic arthropathy. Thirty-one patients (30 males) between 8 and 34 years old (medium age = 20.6 years) were treated with a fixed intra-articular dose of 5 mCi (185 MBq) and divided into two groups: paediatric (13 patients aged up to 18 years, with a medium age of 12.7 years and arthropathy evolution of 7.8 years); and adults (18 patients over 18 years old, with an average age of 24 years and arthropathy evolution of 18.7 years). Clinical assessment before and 1 year after synoviorthesis used the following criteria: subjective (pain according to the visual analogue scale, joint inspection), objective (joint movement through flexion level, pain to palpation and leakage through joint circumference), reduced use of the coagulation factor, number of haemarthroses, and the occurrence of adverse effects. The results were classified as: 1, good (symptom remission of 70–100%); 2, moderate (symptom remission between 40 and 69%); and 3, poor (0–39% symptom remission). Seventy-eight joints were tested: 15 knees, 36 elbows, 24 ankles, 1 shoulder and 2 hips. Early (1–2 h) and late phase scintigraphic imaging (24–72 h) was made after synovectomy. No significant inter-group difference in synovectomy results was observed. The results obtained were good for 75% of elbows, 87.5% of ankles and 40% of knees; reduction in effusions and use of the coagulation factor were, respectively, 78% and 80% for elbows, 82% and 85% for ankles, and 30% and 35% for knees. Four cases of reactive synovitis were observed in the 78 joints tested. Scintigraphy showed homogeneous distribution of the material with no leakage.

The introduction of 153Sm-HA in the treatment of the haemophilic arthropathy is effective for intermediate joints (elbows and ankles), but less so for knees. Moreover, this treatment offers excellent safety and is affordable [3].

The penetration of beta energy from ^{153}samarium (153Sm) (0.8 MeV) is not only suitable for synoviorthesis of intermediate joints, but can improve the radionecrosis effect using higher radioactivity levels. The next study assessed the efficacy of 5 mCi (185 MBq) and 20 mCi (740 MBq) of 153-Sm hydroxyapatite (^{153}Sm-HA) in the knees of haemophilic patients. Thirty-one patients (36 knees, 30 males) were divided into two groups without corticosteroid co-injection: 1 – 14 patients (17 knees) treated with an intra-articular dose of 5 mCi of ^{153}Sm-HA, medium age 23 years; 2 – 17 patients (19 knees), administering 20 mCi of ^{153}Sm-HA, medium age 21.3 years. Evaluation before and 1 year after synoviorthesis used the following points: reduction in the number of effusions and use of the coagulation factor, and increment in joint mobility. The occurrence of side effects was also considered. Early and late-phase scintillations studies were made after synovectomy and no articular immobilisation was recommended. Reduction in effusions and use of the coagulation factor, respectively, were: group 1 – 31.3% and 25%; group 2 – 81.5% and 79%, with $p < 0.001$. No significant increment in knee mobility was observed in either group. Four cases of mild reactive synovitis were observed in each group. Scintigraphy showed homogenous distribution of the radioactive material with no leakage; the material was considered safe by its retention in the joint. A significant increment was observed in the synoviorthesis of haemophilic knees with 20 mCi of ^{153}Sm-HA; the lower penetration of its beta radiation was offset by the improved radiobiological effect when higher radioactivity is used [4].

Another study compared the use of 20 mCi (740 MBq) of ^{153}Sm and 5 mCi (185 MBq) of ^{90}Y, both labelling hydroxyapatite (HA), for knee synoviorthesis in haemophilic patients, 1 year after the procedure. Thirty-three men (36 knees) were studied, divided into two groups: a – injection of 740 MBq of ^{153}Sm-HA: 20 knees of 18 patients, with an average age of 21.4 ± 13.3 years (range: 7–56 years) and medium Pettersson score of 5.3; b – injection of 185 MBq of ^{90}Y-HA: 16 knees of 15 patients, with an average age of 26.3 ± 10.3 (range: 7–51 years) and medium Pettersson score of 6.3. Episodes of haemarthrosis, use of clotting factors and pain intensity were evaluated before and after treatment, as well as improved joint mobility. The occurrence of side effects in the treatment was also considered. The chi-squared, Wilcoxon and Mann–Whitney tests were applied, with a significance level of $p \leq 0.05$. The occurrence of effusions decreased by 65.7% with the use of ^{153}Sm-HA and 82.6% for ^{90}Y-HA, without statistical significance between the groups ($p = 0.632$); pain reduction was 42.5% in group a and 30.7% in group b, again without statistical significance ($p = 0.637$). Increment in joint mobility was not significative for either group. Two cases of mild reactive synovitis were observed in group a and one in group b, which had resolution without medical intervention. Although the beta energy from ^{90}Y is the more appropriate for knee synoviorthesis, the higher radioactivity levels of ^{153}Sm is an alternative in locations that only produce this material [5].

8. Follow-up

This study aimed to evaluate synovectomy with ^{153}Sm-hydroxyapatite (^{153}Sm-HA) in synovitis of the elbows and ankles of haemophilic patients. Synovectomy was performed using

185 MBq of ^{153}Sm-HA in 166 joints (63 ankles and 84 elbows) of 82 haemophilic patients (average age 24.4 years) with follow-up of 12 and 42 months. Arthropathy was characterised by recurrent joint bleeding. Episodes of haemarthrosis, use of clotting factors and pain intensity were evaluated before and after treatment. Scintigraphic analyses and adverse effects were also considered. Statistics used $p \leq 0.05$. The results indicated: (a) reduction in haemarthrosis was 78% and 68%, in elbows and 82% and 72% in ankles; (b) use of clotting factors was 80% and 70% for elbows, and 85% and 75% for ankles; (c) pain intensity was 37% and 34% in elbows, and 61% and 57% in ankles, after 12 and 42 months, respectively. Among the 166 joints studied, three cases of mild reactive synovitis were observed in ankles and four in elbows, with no leakage in any of the cases. In conclusion, the use of ^{153}Sm-HA in elbows and ankles was effective, very safe, minimally invasive and the results showed consistency at follow-up [6].

Another interesting aspect to consider is treatment repetition. We recommend this be done after 1 year, but a minimum interval of 6 months is permitted.

9. Final considerations

The radioactive material (153Sm) was aggregated with hydroxyapatite particles to ensure longer intra-articular retention without arterial-venous or peri-articular lymphatic leakage. When the two are separated, the advantage of the compound is that the carrier (hydroxyapatite) enters the body's metabolism because it is part of the bone matrix.

No escape (lymphatic or vascular) was detected with 153-samarium because when it separates from the carrier (hydroxyapatite), it forms insoluble compounds with the synovial fluid that precipitate in the articulation; this permanence was confirmed by other previous studies [20] and by our controls (see **Figure 3**).

Scintigraphic images obtained after early and late-phase injection showed appropriate intra-articular distribution, as well as the absence of leakage to regional lymph nodes or other organs, or urinary elimination. The ability to obtain good-quality scintigraphic images is an advantage of 153Sm since gamma emission occurs in the energy amplitude of 100 keV. Another possible advantage is the mild reactive synovitis observed in all the studies, resolved without invasive medical intervention, possibly due to low beta energy penetration.

These studies are among the few that evaluate only the therapeutic effect of intra-articular radioactive material, since it is often administered in conjunction with corticosteroids. This creates bias in result analysis because corticosteroids are also used for the same purpose in the treatment of haemophilic arthropathy [11]. This combination has been called into question [13] and is one of the reasons why we chose not to use it when beginning treatment [3], in addition to the lack of information in the literature characterising the nature of its effect as competitive, additive or synergistic.

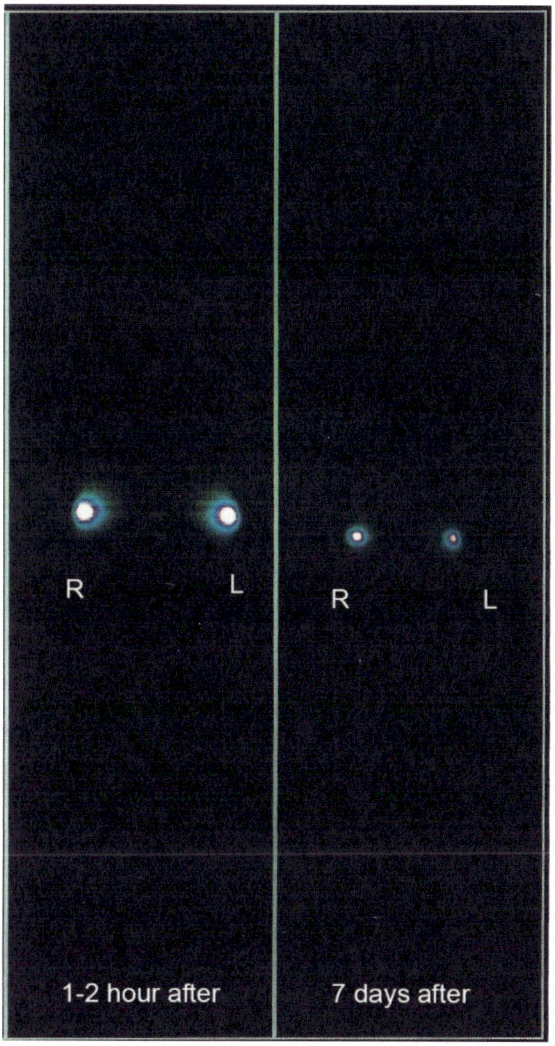

Figure 3. Knee scintigraphy showing good joint distribution and no leakage.

The reduction in haemarthrosis for ankles (82%), elbows (78%) and knees (65.7%) was similar to values recorded in other studies of haemophilic patients with different types of radioactive material. A German revision pointed nine studies between 1982 and 1991 with good and excellent results for radiosynovectomy in 60–80% of the haemophilic arthropathy [9]. The findings presented in our studies also corroborate those summarised by Siegel et al. [25] regarding the benefits of radiosynoviorthesis, with an approximate 75% decrease in effusions, few adverse effects and better quality of life in 75% of the cases. This can be extended to shoulders and hips, which exhibited similar results to intermediate joints. Finally, it can be concluded that 153samarium labelled with hydroxyapatite is a useful tool in the treatment of chronic synovitis in haemophilic patients.

Author details

José Ulisses Manzzini Calegaro

Address all correspondence to: jum.calegaro@gmail.com

Gerência de Ensino e Pesquisa in the Hospital de Base do DF, Nuclear Medicine Unit, Instituto de Radioisótopos de Brasília, Brasil

References

[1] Ahlberg A. Radioactive gold in the treatment of chronic synovial effusion in hemophilia. In: Ana F, Denson R, editors. Hemophilia. Proc. VII Congress World Federation of Hemophilia. Amsterdam, Holland, Teherat Excerpta Medica; 1971. pp. 212-215

[2] Barboza MF, Casiglia MT, Souza AA, Herrerias R, Bambalas E, Santos C, Alves J, Osso JA Jr, Calegaro JUM. 153Sm-HA: Radiopharmaceutical agent for synovectomy. The Quarterly Journal of Nuclear Medicine and Molecular Imaging. 2006;**50**(Suppl 1):46-47

[3] Calegaro JU, Machado J, de Paula JC, de Almeida JSC, Casulari LA. Clinical evaluation after 1 year of 153-samarium hydroxyapatite synovectomy in patients with haemophilic arthropathy. Haemophilia. 2009;**15**:240-246

[4] Calegaro JUM, Machado J, Furtado RG, de Almeida JSC, de Vasconcelos AVP, Barboza MF, de Paula AP. The use of 185 MBq and 740 MBq of 153-samarium hydroxyapatite for knee synovectomy in haemophilia. Haemophilia. 2014;**20**:421-425

[5] Calegaro J, Silveira C, Hage D, Sayago M, de Landa D, Mengatti J, de Paula AP. Comparison of the effect of knee synovectomy in haemophilic patients with ^{153}Sm- and ^{90}Y-labelled hydroxyapatite 1 year after. Haemophilia. 2014;**20**:873-878

[6] Calegaro JUM, Hage DP, Machado J, Sayago M, de Landa DC. Synovectomy using samarium-153 hydroxyaptite in the elbows and ankles of patients with hemophilic arthropathy. World J Nucl Med. 2018;**17**(1): under preparation for issue

[7] Chinol M, Vallabhajosula S, Goldsmith SJ, Klein J, Deutsck KF, Chinen LK, et al. Chemistry and biological behavior of samarium-153 and rhenium-186-labeled hydroxyapatite particles: Potential radiopharmaceuticals for radiation synovitis. Journal of Nuclear Medicine. 1993;**34**:1536-1542

[8] Delbarre F, Dayla J, Menkes CJ, Aignan J, Roucayrol JC. La synoviorthèse par le radioisotopes. Presse Med. 1968;**76**:1045-50

[9] Deutsch E, Brodack JW, Deutsch KF. Radiation synovectomy revisited. European Journal of Nuclear Medicine. 1993;**20**:113-127

[10] Fellinger K, Schmid J. Die locale behandlungder der rheumatischen erkrankungen. Wien Z Inn Med. 1952;**33**:351-363

[11] Fernandez-Palazzi F, Caviglia HA, Salazar JR, Lopez J, Aoun R. Intraarticular dexamethasone in advanced chronic synovitis in hemophilia. Clinical Orthopaedics and Related Research. 1997;**343**:25-29

[12] Gedik GK, Ugur O, Atilla B, Pekmezci M, Yildrim M, Seven B, Varoglu E. Comparison of extra-articular leakage values of radiopharmaceuticals used for radionuclide synovectomy. Annals of Nuclear Medicine. 2006;**20**:183-188

[13] Gedih GK, Ugur O, Atilla B, Dundar S. Is corticosteroid coinjection necessary in radiosynovectomy of patients with hemophilia? Clinical Nuclear Medicine. 2004;**29**:538-542

[14] Heim M, Tiktinski R, Amit Y, Martinowitz U. Yttrium synoviorthesis of the elbow joints in persons with haemophilia. Haemophilia. 2004;**10**:590-592

[15] Jansen NW, Roosendaal G, Bijlisma JW, et al. Degenerated and healthy cartilage are equally vulnerable to blood-induced damage. Annals of the Rheumatic Diseases. 2008; **67**:1468-1473

[16] Klett R, Duille M, Matter HP, Steiner D, Sturz H, Bauer R. Activity leakage and radiation exposure in radiation synovectomy of the knee. Z Rheumatol. 1999;**58**:207-212

[17] Manil L, Voisin P, Aubert B, Guerreau D, Verrier P, Lebègue L, Wargnies JP, Di Paola M, Barbier Y, Chossat F, Menkes CJ, Tébib J, Devaux JY, Kahan A. Physical and bio logical dosimetry in patients undergoing radiosynoviorthesis with erbium-169 and rhenium-186. Nucl Med Commun. 2001;**22**:405-416

[18] Neves M, Waerenborgh P, Patrício L. Palladium-109 and holmium-161. Potential radionuclides for synoviotherapy: Radiation absorbed dose calculations. International Journal of Radiation Applications and Instrumentation. Part A. 1987;**38**:745-749

[19] O'Duffy EK, Oliver FJ, Chatters SJ, et al. Chromosomal analysis of peripheral lymphocytes of patients before and after radiation synovectomy with samarium-153 particulate hydroxyapatite. Rheumatology. 1999;**38**:316-320

[20] O'Duffy ER, Clunie GP, Lui D, Edwards JCN, Ell PJ. Edwards JCN, Ell PJ. Double blind glucocorticoid controlled trial of samarium-153 particulate hydroxyapatite radiation synovectomy for chronic knee synovitis. Ann Rheum Dis. 1999;**58**:554-558

[21] Ofluogo S, Schwameis E, Zenetgruber H, Havlik E, Wanivenhous A, Schweeger I, Weiss K, Sinzinger H, Pirich C. Radiation synovectomy with 166Ho-ferric hydroxide: A first experience. Journal of Nuclear Medicine. 2002;**43**:1489-1149

[22] Rodriguez-Merchan EC, Jimenez-Yuste V, Aznar JA, et al. Joint protection in haemophilia. Haemophilia. 2011;**17**(Suppl. 2):1-23

[23] Schneider P, Farahati J, Reiners C. Radiosynovectomy in rheumatology, orthopedics, and hemophilia. J Nucl Med. 2005;**46**(Suppl 1):48s-54s

[24] Siegel ME, Siegel HJ, Luck JV. Radiosynovectomy's clinical applications and cost effectiveness: A review. Seminars in Nuclear Medicine. 1997;**27**:364-337

[25] Siegel HJ, Luck JV, Siegel ME. Advances in radionuclide therapeutics in orthopedics. J Am Acad Orthop Surg. 2004;**12**:55-64

[26] Teigland JC, Tjonnfjord GE, Evensen SA, Charania B. Synovectomy for haemophilic arthropathy: 6-21 years of follow-up in 16 patients. J Int Med. 1994;**235**:139-243

[27] Vuorela J, Sokka T, Pukkala E, Hannonen P. Does yttrium radiosynovectomy increase the risk of cancer in patients with rheumatoid arthritis? Annals of the Rheumatic Diseases. 2003;**62**:251-253

[28] Winston MA, Bluestone R, Crachiolo A III, Blahd WH. Radioisotope synovectomy with 32P-chromic phosphate-kinetic studies. Journal of Nuclear Medicine. 1973;**14**:886-888

Section 2

Application of Hydroxyapatite: A Synergistic Outlook

Chapter 5

Hydroxyapatite-Based Coating on Biomedical Implant

Wan Sharuzi Wan Harun, Rahil Izzati Mohd Asri,
Abu Bakar Sulong, Saiful Anwar Che Ghani and
Zakri Ghazalli

Additional information is available at the end of the chapter

http://dx.doi.org/10.5772/intechopen.71063

Abstract

The use of metallic biomaterials for replacement of biomedical implants has been traced back from the nineteenth century. These metallic biomaterials have been declared as clinical success as their mechanical properties that satisfy the prerequisite of the human bone. Nevertheless, critical issues of the materials when they are utilised as implants; including the releasing toxic and harmful metal ions through wear and corrosion processes after longer implantation. Besides that, the bonding strength between bone and implants itself is considered weak due to the different components of human bone and metal implants. Hence, developing hydroxyapatite (HAp) coating on metallic biomaterials is expected to overcome the problems faced by biocompatible metallic biomaterials. As far as this, various commercial techniques have been introduced to develop the HAp coating on metallic biomaterials. The techniques are including plasma-spraying method, sol-gel dip-coating, electrochemical deposition and high-velocity suspension plasma-spraying. The formation of HAp coating on metallic biomaterials improved the corrosion resistance together promoting its load-bearing ability and enhanced substrate-coating adhesion.

Keywords: surface coating, biocompatible metals, coating techniques, biomedical applications

1. Introduction

Metals or also known as metallic biomaterials that have been used for medical treatments can be traced back around 20 years. Despite a significant number of metals that are able to be produced in modern industries, there are only three commonly biocompatible metals that are used as biomedical implant materials; stainless steel (316L), cobalt-based alloys, and titanium-based alloys [1–4]. These metallic biomaterials are typically used in orthopaedic practise since they have approval by the United States Food and Drug Administration (FDA) [4].

Since the biomaterials are employed in intimate contact with living tissues, it is important that the materials exhibited biocompatibility characteristics. The requirement of biocompatibility includes all features of bio-device functionalities during the interaction of tissues and cells with the implanted materials [5]. However, there are limitations of metallic biomaterials as an implant; weaknesses in bone-bonding ability and toxic ions released into the human body fluids after longer usage [6, 7]. Different chemical composition between the actual bone and the metal implant is one of the causes of ineffective in bone-bonding ability. Moreover, the metallic implants are also susceptible to corrosion degradation due to the surrounding aggressive body fluids [8].

Consequently, most of the researchers have introduced surface modification by applying bioactive ceramics such as hydroxyapatite ($Ca_{10}(PO_4)_6OH_2$) as a bioactive coating on the metallic implants to the implant to enhance bone-bonding ability [9, 10]. The hydroxyapatite (HAp) is the primary inorganic ingredient of natural bones and has been the most widely used ceramic-based biomaterial for over four decades in medicine and dentistry. It has been proven by many researchers that HAp coating allows a controlled and rapid osseointegration between living bone and the surface of an implant [11, 12].

There are various commercial techniques to deposit the HAp coating on the metal-based biomaterials. In this chapter, four commercially HAp surface coating techniques including plasma spraying, sol-gel dip-coating, electrochemical deposition and high-velocity suspension plasma-spraying (HVSPS) are discussed. The discussion comprises biocompatibility, adhesion strength and corrosion behaviour studies about three aforementioned metallic biomaterials after the surface was coated by HAp.

2. Current issues of metallic biomaterials when applied as an implant

The selection of appropriate biomaterial to be classified as a metal implant material highly depends on its applications. The selected biomaterial should possess several essential characteristics such as excellent biocompatibility, osseointegration, high corrosion and wear resistance, suitable mechanical properties, ductility and high hardness.

2.1. Biocompatibility and osseointegration

The biocompatibility properties are defined as the ability of a material to be used in intimate contact physically and chemically with living tissues of a real bone without causing any adverse effects. Intuitively, it is necessary to confirm that there are no negative issues befell to metal implanted devices and surrounding living tissues since the materials are innately compatible with living cells and tissues [13].

Osseointegration denotes to a direct structural and functional connection between ordered, living bone and the surface of a load-carrying implant. It involves the process of new bone production and bone healing. Therefore, it is essential for an implant to have an appropriate surface to integrate well with surrounding bone. Surface chemistry, surface roughness and surface topography are the factors that vital for good osseointegration [14, 15].

2.2. Corrosion and metal ions release

Among critical issues and challenges of the medical implant are facing is the failure of an implant due to the corrosion aggressiveness. Consequently, a metal that performs well outside the human body may suffer a severe corrosivity reaction in the body as the environment is physically and chemically different from ambient. Due to that fact, all of the corrosion-resistant metallic implants reacted to an acidic environment and began to corrode when diagnosed for a long time in the human body. Most researchers have claimed that active implants corrosivity process rouse after 12–15 years of implantation period [6, 16].

3. Commercial techniques for hydroxyapatite-based coating onto metallic implant

The surface coating application offers the possibility of modifying the surface properties of implant devices to achieve improvements in biocompatibility, reliability, and performance. Therefore, most researchers have reported excellent studies of HAp coating onto various metallic implants specifically related to their biocompatibility and corrosion behaviour. Nowadays, different HAp deposition techniques have been carried out to overcome the biocompatibility, and corrosion issues arose from the metallic biomaterials [12, 17]. These deposition techniques include plasma spraying, sol-gel technique, electrochemical deposition and High-Velocity Suspension Plasma-spraying.

3.1. Plasma-spraying technique

Plasma spray is one of the popularly used methods used to deposit biocompatible HAp coating onto metallic implants [18, 19]. In these recent years, this approach is highly utilised for dental and orthopaedic implants. The indirect method of plasma spray applies melting and spraying onto the surface by a method an electric arc. The process involves heating the dry powder feedstock by thermal plasma jet. Then, the thermal plasma jet accelerates and impacts the feedstock towards the substrate. The powder feedstock is flattened in the form of lamellae. Plasma spraying can be carried out under vacuum, controlled atmospheres, or in an ambient atmosphere. Air or vacuum spraying is one of the plasma-assisted depositions, which is very popular compared to other methods. The coatings applied by plasma spray can have relatively good mechanical properties. The relative temperatures in the jet are 10,000 K, 12,000 K or as high as 30,000 K, intensely declining with the nozzle's distance [20]. Practically all the materials are melted and propelled towards a substrate.

Development of the coated layer on a titanium (Ti) alloy surface with hydroxyapatite powder for 10 s shows better apatite adhesion, strong adhesion between implant and bone, and enhanced osteoconductivity [21]. The properties of HAp coating are mainly determined by the thickness of the coating layer. The thickness of HAp coating obtained on the Ti6Al4V alloy by the air plasma spray (APS) was about 150 μm thickness. This range of coating's thickness significantly diminishes the fatigue strength while the range between 25 and 100 μm thickness does not show such effect [22]. The reasons for the reduction in fatigue strength might be due to the intrinsic stresses that happened during spraying, coating cracks, and most significantly stresses discharged during spraying. The difference in the stiffness of the metal substrate and coating is also significant.

The coating deposition by a suspension plasma spray (SPS) obtained a relatively thin coating layer 5–50 μm as compared to other plasma-spray techniques and only could be achieved by dry powder processing [23]. The dry powder particles used for the SPS having diameters ranging from a few submicrometer to a few micrometres [24]. The thicknesses of APS coatings are in the range 200–300 μm and quite porous. The coating thickness depends on the composition of plasma gas used which is $Ar/H_2/N_2/He$, plasma gun input power, gas flow rate, powder feeding rate and characteristics of feed materials, and spray stand-off distance, which are frequently varied [25, 26].

Furthermore, the structure and bonding properties of HAp coating on metallic biomaterials can be improved by using heat treatment process. Annealing process transformed a partial amorphous coating into a crystalline layer [22]. The mostly higher crystallinity of the coating layer was supposed to have excellent adhesion characteristics. Annealing at a higher temperature such as at 700°C for 1 h could enhance the coating purity, hydroxyl group and crystallinity degree. However, the high spraying power values can cause a lowering of the adhesion strength between the coating and substrate due to the higher content of amorphous HAp [27]. Based on the observations of the annealing process of HAp coated at 1100°C under vacuum condition, the secondary β phase formed while hydroxyl groups are diminished. The higher the temperature of the annealing process, the greater the formation of the compound oxide of Ti and Ca with the characteristic metallic Ti disappeared [28].

An introduction a coupling agent through chemical bonding can enhance adhesion strength between the HAp coating and metallic implants [29]. The addition of Ti to the HAp improved the bonding strength of the coating significantly [30]. The bonding strength was increased from 14.5 to 17.3 MPa as the composition of the reinforced coating was between 20 and 60 wt.% Ti. The increment in the Ti content could cause better adhesion of the coating layer to the substrate for further enhancement. According to Ref. [31] proposed HAp reinforced with 10 wt% ($80Al_2O_3$-$20TiO_2$) on the Ti6Al4V alloy. This solution enhances the adhesion strength to above 32 MPa.

3.2. Sol-gel dip-coating method

Recently, a combination of sol-gel preparation and dip-coating method are extensively employed for a coating on a metallic biomaterial. The method is one of the coating methods used for enhancement of adhesion strength [32, 33]. A calcium phosphate (CaP) precursors are the most important solutes for sol preparation. The CaP precursor is the combination of calcium (calcium nitrate) and phosphorus (phosphorus pentoxide or triethyl phosphate). Normally, there are two solvents will be mixed with the CaP precursors. Most often water and ethanol are used as a solvent for the sol preparation [34, 35].

The dip-coating is a method which includes three steps: (i) dipping, (ii) withdrawing, and (iii) drying. This technique offers various advantages such as low-cost set-up, process simplicity, uniformity of deposition, low processing temperature, and the ability to coat irregular shapes and patterns [36, 37]. The substrate is dipped and withdrawn from the solution at a fixed speed. Therefore, the coating's thickness is in good control without producing waste [38, 39].

Additionally, the coating amount and the layer thickness can also be controlled by altering the frequency of suspension and the number of dippings. HAp coating via sol-gel dip-coating technique can obtain homogenous coating and the coating thickness in the range 0.05–15 mm [33, 40]. The coating thickness varies according to the viscosity of the sol-gel used [12]. A lower annealing temperature used for sol-gel dip-coating process can produce adhesive thin coating layer without severe cracking. An extremely high temperature (6000–10,000°C) is applied in plasma-spray deposition can decompose the HAp properties into tricalcium phosphate, tetra-calcium phosphate, calcium oxide (CaO), and others amorphous phases [41, 42]. When increasing annealing temperature from 375 to 500°C, the adhesion strength between HAp coating and the substrate increases [43, 44].

Latterly, several modifications of sol-gel dip-coating method are developed to enhance the quality of coating surface. A poly ε-caprolactone (PCL) was applied to HAp to promote osseointegration by observing the pores formation on a surface level [45]. The addition of PCL on HAp onto Ti6Al4V substrate was reported as a good grouping owing to a large thickness of the coating, around 184 μm. There was no crack formation on the coating surfaces, and the most significant results revealed that the adhesion between the coating and the substrate was improved. The absence of cracks on the coating surface was reflected necessarily. This is an effective prevention of wear and corrosion for the substrate. Hence, the amount of releasing metal ions into surrounding (body fluid) can be minimised as the coated metal exhibit better corrosion resistance [46–48].

Heat treatment of thin and loosely packed coated substrate is often required to densify the coating layer and to increase the adhesion strength between the substrate and coating [47, 49, 50]. The high temperature is applied to cure the coated substrates to improve the adhesion strength between coating and substrate, and to achieve apatite structures inside the applied coating layers [51, 52]. However, the curing temperatures have been implemented below the melting point of the materials to prevent upsetting the surface integrity of the substrates. It has been indicating that the development of <1 μm thickness of HAp coating on 316L stainless steel was suited as the substrate also exposed to annealed temperatures of around 375–400°C [43]. The bonding strength of the as-produced coatings was about 44 MPa, which indicates good adhesion. For the presently investigated HAp/316L stainless steel system, the interlocking component of adhesion was maximised through surface roughening.

Lately, a modification of the sol-gel dip-coating has been proposed. TiO_2/HAp bi-layer coating and TiO_2/HAp composite coating were introduced into 316L stainless steel (316L SS) [53]. The two types of the coating were compared, and TiO_2/HAp exhibited better structural features and biocompatible properties due to the proper attachments of stem cells onto the surface, proliferated, and presented a polygonal morphology different from the fibroblastic-like morphology found on 316L SS.

The combination of sol-gel and dip-coating method have been classified as uncomplicated, inexpensive, and sustainable coating technique for coating the metal-based substrate that to be used as implants. In comparison to the natural precipitation approaches, this method can coats the complex shapes or design efficiently. Also, shortened the processing times and relatively low temperatures is applied by this method to cover with the HAp layers on metallic substrates.

3.3. Electrochemical deposition

Electrochemical deposition is one of the commercial coating methods for biomedical implants [54]. Anodic or cathodic systems frequently conducted by the electrodeposition process. In this process, anodic deposition alone is inadequate to produce small feature size materials on the substrates. Regarding this, cathodic deposition has unique advantages for modern and medical applications [55–57]. Through this method, two regular procedures are applied for the coating preparation: (1) the electrophoretic process (EPD) and (2) the electrolytic procedure (ELD). EPD is the process that provides the utilisation of suspensions of ceramic particles while the ELD is the process of formation of metal salts from solutions. The electrochemical deposition is extensively employed for coating on a titanium substrate. The subsequent filtrate is used as the electrolyte once the CaP proportion dissolved in distilled water [58, 59]. The procedure is performed from watery arrangements like those appropriated as a part of the wet substance deposition. Interestingly, graphite and also platinum has been engaged as the standard reference electrode for anodic material.

One of a kind favourable circumstances in HAp coating deposition process conducted through the electrochemical technique is the ability to form a uniform coating and the coating process quick [60]. The procedure can be performed at moderately low temperature [61]. Electrophoresis process can produce impregnated ceramic particle towards a porous substrate and composite consolidation. Besides, a significant aspect of sintering behaviour greatly depends on the state of agglomeration of ceramic powders. The lower the sintering temperature, the more densify the close-packed of the fine particle and further leads to the formation of agglomerate-free structures [62]. The pre-sedimentation process can separate the aggregates [63]. Besides, defect areas could drive a higher rate of deposition, bringing about the uniformity of the deposit materials and better packing assembly of materials. The higher rate and better deposition layer are due to the insulating behaviour of the deposition.

The electrochemical deposition process can form a homogeneous coating layer which enhances the adhesion strength between the coating layer and implant surface [64]. The HAp coating on CoCrMo metal implant with the thickness of 200 nm can hold the coating quality of around 17.5 MPa which has been considered as the base prerequisite for the minimum adhesion quality of HAp deposition on metallic biomaterials [65, 66]. However, the HAp coated substrate was deposited at 10 mA/cm^2 and annealed for 1 h at 500°C showed the thickness of the coating is approximately 18.6 μm revealed stronger adhesion strength (106.3 MPa) of HAp coating [67]. The electrochemical deposition of HAp on metal substrates used common strategies to diminish their debasement; unfortunately, it contains abandon of it onto the coating surface [68]. Hydroxide (OH−) particles are created at the substrate (cathode) surface with the electric current crossed the electrodes as they immersed in an electrolyte during electrochemical deposition process [69]. The condition occurs due to the electrochemical response effect that is typically significant as the system insignificantly response towards the water, in which leads to important in the arrangement of a lot of hydrogen gas [27, 70]. Development of the hydrogen gas air pockets on the surface of the substrate may rapidly occur and thus results in the decreasing of the nucleation and presence of calcium phosphate. In this manner, it may prompt the arrangement of non-uniform coating [71, 72]. To conquer the defects, the execution of HAp coating ought to tackled and enhanced higher current thickness.

A few modifications are recommended to adjust the direct current electrochemical deposition technique that should be more reasonable in the accompanying approach. H_2O_2 was added to replace the H_2O during the deposition process, thus brings down the current deposition method. Replacement of H_2O_2 will able to modify the entire part of the system of electrochemical response [73]. The impacts of H_2 development might be evacuated due to the expansion of peroxide. Therefore, the thick and uniform coating might be shaped [61]. The increased adhesion and crystallinity of the HAp coating were achieved by pulsed current electrodeposition method at lower current density with longer pulse off time. The results of pulsed electrodeposition show that the relaxation time of the pulse is beneficial for the growth of HAP because it allows the diffusion of ions from bulk solution to the surface of the electrode and thus lowers the concentration polarisation in the next pulse on time. Besides, by applying galvanostatic pulse electrodeposition to HAp coating on metal implants showed improvement in adhesion strength of HAp coating and metal implant due to pulsed current densities [74].

The previous research has confirmed that by deposition of HAp coating onto metallic biomaterial showed the improvement of corrosion performance [75]. Moreover, the coatings have significantly changed by forming new apatite crystal after 7 days immersion in SBF solution [76]. These indicate that HAp has bioactivity and biocompatibility properties which can provide improvement between tissues and metal implants.

An anodization process has been introduced as pre-treatment for electrochemical deposition [77]. The purpose of anodization is to support developing mechanical interlocks flanked by the metal substrate and HAp coating [78, 79]. Without post-treatment, homogenous and pure HAp coating can be accomplished through the anodising process. It was stated by He et al. [80] that the Al coating on Ti substrate with anodization and hydrothermal treatment. The results from the holes of anodised aluminium oxide (Al_2O_3) within the coating deposition, there is a growth of CaP. Besides, Yang et al. [81] have conducted anodic oxidation treatment for bioactive Ti metal. Even though the electrochemical and pre-treatment process has become vital, the studies on HAp coating deposited by an electrochemical method on the porous anodised Ti substrate still in progress stage [82].

Nowadays, a few methods have been presented to enhance mechanical properties of the implants. One of those methods is through reinforcing materials such as zirconia oxide (ZrO_2), carbon nanotubes (CNTs), and titanium oxide (TiO_2) [11, 83–87]. On the other hand, several reported regarding the HAp-based coating showed the enhancement in adhesion strength approaching 70% greater compared to pure HAp coating [88]. HAp coating with the addition of single-walled nanotubes (SWNT) managed to get homogenous, high crystallinity and crack-free coatings formation. Additionally, the adhesion strength of the coating and substrate after introducing SWNT is approaching from 15.3 to 25.7 MPa [88].

In correlation to a single layer coating of HAp, the result of a double layer of HAp coating showed uniformity with good adhesion strength [89]. Furthermore, the formation of the oxide layer as an intermediate layer between the substrate and coating helps to maintain the diffusion of harmful impurities from the substrate towards the coating surface to avoid decomposition of HAp [90]. By applying high-temperature annealing or sintering in the formation of a uniform and denser CaP coating post-electrodeposition [91], a superior adhesion behaviour of coated layer can be formed. Albayrak et al. [92] have reported the same technique used by Yuan and Golden [89]. Titanium oxide (TiO_2) was introduced as an oxide layer on the Ti6Al4V substrate prior HAp coating. The coated substrates with the presence of TiO_2 had the thickness about 30 μm and were soaked for 1 min with different voltages as 10, 20, and 50 V. With decreasing the voltage value, the result showed an increment of adhesion strength. Comparison of the adhesion strength between electrodeposition methods was listed in **Table 1**.

In conclusion, HAp coatings conducted via electrochemical deposition technique are formed progressively by nucleation and growth processes and lead to form a uniform structure. The electrochemical deposition technique can form a broad range of coating thickness. Also, the electrochemical deposition process decreases the corrosion behaviour of the substrate through the coating. Consistently, sintering procedure enhances densification, bonding and adhesion behaviours of the coating. An interlayer between the substrate and the coating has been introduced to overwhelm the issue of HAp decomposition. Therefore, the electrochemical deposition technique can be one of great guarantee of the future edition for metallic biomaterials.

Composition	Thickness (μm)	Adhesion strength (MPa)	Reference
CoCrMo + HAp	0.2	17.5	[66]
1. Ti6Al4V + HAp (flake-shaped)	10	6.8	[83]
2. Ti6Al4V + HAp (spherical)	10	10.7	
3. Ti6Al4V + sHAp/CNT-Ti	10	10.6	
4. Ti6Al4V + HAp (needle-shaped)	10	8.5	
1. Ti + HAp (without oxidation)	3	5.0	[82]
2. Ti + HAp (with oxidation)	3	7.3	
1. Ti + HAp	10	15.3	[88]
2. Ti + SWNTs/HAp	10	25.7	
1. Ti6Al4V + HAp	30	13.8	[92]
2. Ti6Al4V + TiO_2 (10 V)/HAp	30	21.0	
3. Ti6Al4V + TiO_2 (20 V)/HAp	30	13.1	
4. Ti6Al4V + TiO_2 (50 V)/HAp	30	11.9	

Table 1. Values of adhesion strength of HAp coatings deposited by electrodeposition process [33].

3.4. High-velocity suspension plasma-spraying

Currently, High-Velocity Oxygen-Fuel (HVOF) flame spraying method has been advanced. The method gives promising results regarding allowing the formation of suspension spraying layer [93–95]. By introducing the axial powder injection, the new high-velocity suspension flame spraying (HVSFS) process typically would be able to resolve the injection complications [96–98]. Regarding this, the highest velocity of the particle would be able to produce better coating protection with low porosity. This innovative suspension thermal spray technique is ideally becoming the most in-demand technique in depositing a thin layer of coating on the substrate [99, 100]. In comparison to the other method such as electrophoretic coating for pure HAp, the coating does not require any heat treatment or post-deposition for consolidation. Furthermore, the method features relatively lower processing cost with high and efficient productivity [101, 102]. As compared to conventional dry powders, the suspension based feedstock could ignite more flexibility in creating new composite materials by altering the material compositions in which controlling the primary particle morphology [103, 104]. Furthermore, a fine powder particle either in micro- or nano-sized particles could be fabricated by thermal spray community. The method enables direct delivery of the particles into the gas or plasma jet. Direct processing of fine particles dispersed in liquid solvent significantly yield smaller lamella size of the coating layer that depends on the standard parameters of spray powder processing [105].

HVSFS technique could produce high-quality and low-thickness coatings especially when the layer thickness is below 50 μm [96, 106]. The development of the system does indeed fill in the gap between conventional thermal spraying and thin-film technologies (PVD, CVD). The thinner coatings produced from this technique usually contain less residual stress with minimising risks of delamination [107, 108]. The coated properties especially the one involves bonding strength between the substrate and coated layer produced from HVSFS techniques tends to be affected severely due to the effect of processing parameters such as gas flow, air-fuel ratio spray distance, and electric arc current. As reported by Gadow et al., [106], bioceramic coatings could be based on dry spray HAp powder used for HVOF and APS nanoscale and HAp suspension (water-based) for HVSFS.

By introducing different suspension solution medium such as diethylene glycol (DEG) as a substitute of water suspension, the result of the adhesion strength of the HAp coatings is enhanced and supposed to be superior [106]. The DEG-based solution increases the adhesion strength compared to the water-based suspension. In fact, DEG-based suspension offers many advances properties such as low interlayer porosity with denser coatings. The resulting condition is due to the higher adhesion strength. The maximum adhesion strength is around 25 N/mm^2 as reported for HVSFS-HAp coating [106]. Additional of TiO_2 layer acts as a thermal insulating layer onto commercially pure Ti slows down heat extraction from the deposited material to the substrate [109, 110]. The HVSFS deposited HAp coatings are typically dense with 27–37 μm in thickness and some transverse micro-cracks. Even the crystallinity characteristic is of between 10 and 70%, depending on the deposition parameters and the amount of a TiO_2 concentration. The adhesion strength between the metal substrate and HAp coating enhanced due to the presence of TiO_2 layer as shown in **Figure 1**.

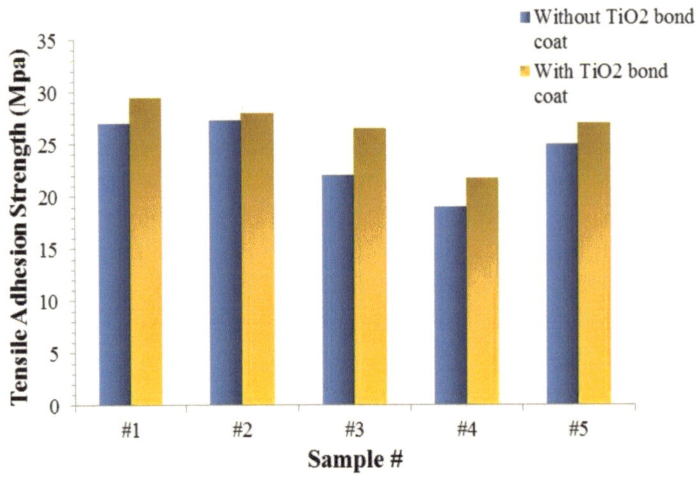

Figure 1. Adhesion strength of the HVSFS-deposited HAp coatings [110].

The APS, HVOF and HVSFS methods are extensively practised for HAp coating processes especially the one involves Ti metal as substrates [111, 112]. The processing parameters such as gas flow, air-fuel ratio, electric arc current and spray distance are the primary factor in determining the coating properties performance concerning the adhesion strength between substrate and coating of these thermal spraying techniques [113, 114]. The increment of the flow rate of oxygen enhances the behaviour of the coating composite. Besides that, fuel flow rate also plays a significant role in influencing the coating performance. Increasing oxygen flow rate along with fuel flow rate leads to higher adhesion strength. Other than that, reducing the spray distance also brings to stronger adhesion strength [98, 104, 115].

4. Summary

The choice of metallic biomaterials such as 316L stainless steel, cobalt-based alloy, titanium and its alloys will continue to be used extensively in the medical field as medical implants due to their excellent mechanical properties and adaptability within the physiological environment. Currently, a major issue of metallic implants is the failure to perfect pair to the local tissue environment in the human body. This inharmonious is due to the different chemical compositions between metallic implants and human bone. The surface modification of the metal-based materials via four common coating techniques namely plasma-spray, sol-gel, dip-coating, electrochemical deposition and high-velocity suspension plasma-spraying (HVSPS) was introduced to enhance bioactivity, to prevent wear and corrosion and to control harmful metal ions released into the body. It is proven that the surface modification via coating can succeed the limitation of the metallic biomaterials.

Acknowledgements

Authors would like to acknowledge the kind assistantship from the Green Research for Advanced Materials Laboratory (GRAMsLab) and Human Engineering Group (HEG), Universiti Malaysia Pahang (UMP). This work was supported by the Qatar National Research Fund NPRP88762375 (UIC161504), internal grant of Universiti Malaysia Pahang RDU140354, RDU160337, and the support of Research Acculturation Grant Scheme provided by the Ministry of Higher Education, Malaysia RDU151404.

Author details

Wan Sharuzi Wan Harun[1]*, Rahil Izzati Mohd Asri[2], Abu Bakar Sulong[3], Saiful Anwar Che Ghani[1] and Zakri Ghazalli[1]

*Address all correspondence to: sharuzi@ump.edu.my

1 Green Research of Advanced Materials Laboratory (Grams Lab), Human Engineering Group, Faculty of Mechanical Engineering, Universiti Malaysia Pahang, Pahang, Malaysia

2 Institute of Postgraduate Studies, Universiti Malaysia Pahang, Lebuhraya Tun Razak, Pahang, Malaysia

3 Department of Mechanical and Material Engineering, Universiti Kebangsaan Malaysia, Selangor, Malaysia

References

[1] Niinomi M. Recent metallic materials for biomedical applications. Metallurgical and Materials Transactions A. 2002;**33**:477-486

[2] Niinomi M, Nakai M, Hieda J. Development of new metallic alloys for biomedical applications. Acta Biomaterialia. 2012;**8**:3888-3903

[3] Breme H, Biehl V, Reger N, Gawalt E. A metallic biomaterials: Introduction. In: Handbook of Biomaterial Properties. Springer; New York. 2016. p. 151-158

[4] Chen Q, Thouas GA. Metallic implant biomaterials. Materials Science and Engineering: R: Reports. 2015;**87**:1-57

[5] Freire WP, Fook MVL, Barbosa EF, dos S Araújo C, Barbosa RC, Pinheiro ÍM. Biocompatibility of dental restorative materials. Materials Science Forum. 2015;**8**:19-25

[6] Asri RIM, Harun WSW, Samykano M, Lah NAC, Ghani SAC, Tarlochan F, et al. Corrosion and surface modification on biocompatible metals: A review. Materials Science and Engineering: C. 2017;**77**:1261-1274

[7] Espallargas N, Torres C, Muñoz AI. A metal ion release study of CoCrMo exposed to corrosion and tribocorrosion conditions in simulated body fluids. Wear. 2015;**332-333**:669-678

[8] Antunes RA, de Oliveira MCL. Corrosion fatigue of biomedical metallic alloys: Mechanisms and mitigation. Acta Biomaterialia. 2012;**8**:937-962

[9] Dorozhkin SV. Calcium orthophosphate deposits: Preparation, properties and biomedical applications. Materials Science and Engineering: C. 2015;**55**:272-326

[10] Luo L, Jiang Z, Wei D, He X. Surface modification of titanium and its alloys for biomedical application. Advance Materials Research. 2014;**887-888**:1115-1120

[11] Ayu HM, Izman S, Daud R, Krishnamurithy G, Shah A, Tomadi SH, et al. Surface modification on CoCrMo alloy to improve the adhesion strength of hydroxyapatite coating. Procedia Engineering. 2017;**184**:399-408

[12] Ben-Nissan B, Choi AH, Roest R, Latella BA, Bendavid A. 2-Adhesion of hydroxyapatite on titanium medical implants. In: Mucalo M, editor. Hydroxyapatite (Hap) for Biomedical Applications. Woodhead Publishing; 2015. p. 21-51

[13] Hussein KH, Park K-M, Kang K-S, Woo H-M. Biocompatibility evaluation of tissue-engineered decellularized scaffolds for biomedical application. Materials Science and Engineering: C. 2016;**67**:766-778

[14] Mahjoubi H, Buck E, Manimunda P, Farivar R, Chromik R, Murshed M, et al. Surface phosphonation enhances hydroxyapatite coating adhesion on polyetheretherketone and its osseointegration potential. Acta Biomaterialia. 2017;**47**:149-158

[15] Wang D, Wu G, Lin X, Liu Y. Coatings for osseointegration of metallic biomaterials. Surface Coating and Modification of Metallic Biomaterials. 2015;**11**:345

[16] Fonseca-García A, Pérez-Alvarez J, Barrera CC, Medina JC, Almaguer-Flores A, Sánchez RB, et al. The effect of simulated inflammatory conditions on the surface properties of titanium and stainless steel and their importance as biomaterials. Materials Science and Engineering: C. 2016;**66**:119-129

[17] Catauro M, Papale F, Sapio L, Naviglio S. Biological influence of Ca/P ratio on calcium phosphate coatings by sol-gel processing. Materials Science and Engineering: C. 2016;**65**:188-193

[18] Liu C, Zhang J, Zhang S, Wang P, Lian Y, Deng S, et al. Preparation and properties of ceramic coatings by cathode plasma electrolytic deposition on titanium alloy. Surface and Coatings Technology. 2017;**327**:714

[19] Heimann RB, Lehmann HD. Deposition, structure, properties and biological function of plasma-sprayed bioceramic coatings. Bioceramic Coatings for Medical Implants. 2015;**6**:253-308

[20] Surmenev RA. A review of plasma-assisted methods for calcium phosphate-based coatings fabrication. Surface and Coatings Technology. 2012;**206**:2035-2056

[21] Liu X, Poon RW, Kwok SC, Chu PK, Ding C. Plasma surface modification of titanium for hard tissue replacements. Surface and Coatings Technology. 2004;**186**:227-233

[22] Lynn A, DuQuesnay D. Hydroxyapatite-coated Ti–6Al–4V: Part 1: The effect of coating thickness on mechanical fatigue behaviour. Biomaterials. 2002;**23**:1937-1946

[23] Gross KA, Saber-Samandari S. Revealing mechanical properties of a suspension plasma sprayed coating with nanoindentation. Surface and Coatings Technology. 2009;**203**:2995-2999

[24] Kozerski S, Pawlowski L, Jaworski R, Roudet F, Petit F. Two zones microstructure of suspension plasma sprayed hydroxyapatite coatings. Surface and Coatings Technology. 2010;**204**:1380-1387

[25] Tang Q, Brooks R, Rushton N, Best S. Production and characterization of HA and SiHA coatings. Journal of Materials Science: Materials in Medicine. 2010;**21**:173-181

[26] Yang C-Y, Lee T-M, Lu Y-Z, Yang C-W, Lui T-S, Kuo A, et al. The influence of plasma-spraying parameters on the characteristics of fluorapatite coatings. Journal of Medical and Biological Engineering. 2010;**30**:91-98

[27] Sobieszczyk S, Zieliński A. Coatings in arthroplasty: Review paper. Advances in Materials Science. 2008;**8**:35-54

[28] Yang Y, Kim K-H, Agrawal CM, Ong JL. Interaction of hydroxyapatite–titanium at elevated temperature in vacuum environment. Biomaterials. 2004;**25**:2927-2932

[29] Yang Y, Kim K-H, Ong JL. A review on calcium phosphate coatings produced using a sputtering process—An alternative to plasma spraying. Biomaterials. 2005;**26**:327-337

[30] Zheng X, Huang M, Ding C. Bond strength of plasma-sprayed hydroxyapatite/Ti composite coatings. Biomaterials. 2000;**21**:841-849

[31] Singh G, Singh S, Prakash S. Surface characterization of plasma sprayed pure and reinforced hydroxyapatite coating on Ti6Al4V alloy. Surface and Coatings Technology. 2011;**205**:4814-4820

[32] Zhang JX, Guan RF, Zhang XP. Synthesis and characterization of sol-gel hydroxyapatite coatings deposited on porous NiTi alloys. Journal of Alloys and Compounds. 2011;**509**:4643-4648

[33] Asri R, Harun W, Hassan M, Ghani S, Buyong Z. A review of hydroxyapatite-based coating techniques: Sol-gel and electrochemical depositions on biocompatible metals. Journal of the Mechanical Behavior of Biomedical Materials. 2016;**57**:95-108

[34] Costa DO, Dixon SJ, Rizkalla AS. One-and three-dimensional growth of hydroxyapatite nanowires during sol–gel–hydrothermal synthesis. ACS Applied Materials & Interfaces. 2012;**4**:1490-1499

[35] Alves Cardoso D, Jansen J, Leeuwenburgh SG. Synthesis and application of nanostructured calcium phosphate ceramics for bone regeneration. Journal of Biomedical Materials Research Part B: Applied Biomaterials. 2012;**100**:2316-2326

[36] Guo L, Li H. Fabrication and characterization of thin nano-hydroxyapatite coatings on titanium. Surface and Coatings Technology. 2004;**185**:268-274

[37] Motealleh A, Eqtesadi S, Perera FH, Pajares A, Guiberteau F, Miranda P. Understanding the role of dip-coating process parameters in the mechanical performance of polymer-coated bioglass robocast scaffolds. Journal of the Mechanical Behavior of Biomedical Materials. 2016;**64**:253-261

[38] Faustini M, Louis B, Albouy PA, Kuemmel M, Grosso D. Preparation of sol−gel films by dip-coating in extreme conditions. The Journal of Physical Chemistry C. 2010;**114**:7637-7645

[39] Yuan J, Zhao K, Cai T, Gao Z, Yang L, He D. One-step dip-coating of uniform γ-Al_2O_3 layers on cordierite honeycombs and its environmental applications. Ceramics International. 2016;**42**:14384-14390

[40] Mohseni E, Zalnezhad E, Bushroa AR. Comparative investigation on the adhesion of hydroxyapatite coating on Ti–6Al–4V implant: A review paper. International Journal of Adhesion and Adhesives. 2014;**48**:238-257

[41] Liu D, Chou H, Wu J. Plasma-sprayed hydroxyapatite coating: Effect of different calcium phosphate ceramics. Journal of Materials Science: Materials in Medicine. 1994;**5**:147-153

[42] Carayon M, Lacout J. Study of the Ca/P atomic ratio of the amorphous phase in plasma-sprayed hydroxyapatite coatings. Journal of Solid State Chemistry. 2003;**172**:339-350

[43] Liu D-M, Yang Q, Troczynski T. Sol–gel hydroxyapatite coatings on stainless steel substrates. Biomaterials. 2002;**23**:691-698

[44] Sidane D, Khireddine H, Yala S, Ziani S, Bir F, Chicot D. Morphological and mechanical properties of hydroxyapatite bilayer coatings deposited on 316L SS by sol–gel method. Metallurgical and Materials Transactions B. 2015;**46**:2340-2347

[45] Mohd Yusoff MF, Abdul Kadir MR, Iqbal N, Hassan MA, Hussain R. Dipcoating of poly (ε-caprolactone)/hydroxyapatite composite coating on Ti6Al4V for enhanced corrosion protection. Surface and Coatings Technology. 2014;**245**:102-107

[46] Hornberger H, Virtanen S, Boccaccini A. Biomedical coatings on magnesium alloys—A review. Acta Biomaterialia. 2012;**8**:2442-2455

[47] Catauro M, Bollino F, Papale F, Giovanardi R, Veronesi P. Corrosion behavior and mechanical properties of bioactive sol-gel coatings on titanium implants. Materials Science and Engineering: C. 2014;**43**:375-382

[48] Alemón B, Flores M, Ramírez W, Huegel JC, Broitman E. Tribocorrosion behavior and ions release of CoCrMo alloy coated with a TiAlVCN/CNx multilayer in simulated body fluid plus bovine serum albumin. Tribology International. 2015;**81**:159-168

[49] Xiao XF, Liu RF. Effect of suspension stability on electrophoretic deposition of hydroxyapatite coatings. Materials Letters. 2006;**60**:2627-2632

[50] Balamurugan A, Balossier G, Kannan S, Michel J, Rajeswari S. In vitro biological, chemical and electrochemical evaluation of titania reinforced hydroxyapatite sol–gel coatings on surgical grade 316L SS. Materials Science and Engineering: C. 2007;**27**:162-171

[51] Xia W, Lindahl C, Lausmaa J, Borchardt P, Ballo A, Thomsen P, et al. Biomineralized strontium-substituted apatite/titanium dioxide coating on titanium surfaces. Acta Biomaterialia. 2010;**6**:1591-1600

[52] Dinda GP, Shin J, Mazumder J. Pulsed laser deposition of hydroxyapatite thin films on Ti–6Al–4V: Effect of heat treatment on structure and properties. Acta Biomaterialia. 2009;**5**:1821-1830

[53] Sidane D, Rammal H, Beljebbar A, Gangloff SC, Chicot D, Velard F, et al. Biocompatibility of sol-gel hydroxyapatite-titania composite and bilayer coatings. Materials Science and Engineering: C. 2017;**72**:650-658

[54] Dickerson JH, Boccaccini AR. Electrophoretic Deposition of Nanomaterials. Vol. 50. Springer Science & Business Media; 2012

[55] Zhitomirsky I. Cathodic electrodeposition of ceramic and organoceramic materials. Fundamental aspects. Advances in Colloid and Interface Science. 2002;**97**:279-317

[56] He D-H, Wang P, Liu P, Liu X-K, Ma F-C, Zhao J. HA coating fabricated by electrochemical deposition on modified Ti6Al4V alloy. Surface and Coatings Technology. 2016;**301**:6-12

[57] Eliaz N, Shmueli S, Shur I, Benayahu D, Aronov D, Rosenman G. The effect of surface treatment on the surface texture and contact angle of electrochemically deposited hydroxyapatite coating and on its interaction with bone-forming cells. Acta Biomaterialia. 2009;**5**:3178-3191

[58] Yajing Y, Qiongqiong D, Yong H, Han S, Pang X. Magnesium substituted hydroxyapatite coating on titanium with nanotublar TiO_2 intermediate layer via electrochemical deposition. Applied Surface Science. 2014;**305**:77-85

[59] Bakin B, Delice TK, Tiric U, Birlik I, Azem FA. Bioactivity and corrosion properties of magnesium-substituted CaP coatings produced via electrochemical deposition. Surface and Coatings Technology. 2016;**301**:29-35

[60] Zhao X, Yang L, Zuo Y, Xiong J. Hydroxyapatite coatings on titanium prepared by electrodeposition in a modified simulated body fluid. Chinese Journal of Chemical Engineering. 2009;**17**:667-671

[61] Zhao X, Hu T, Li H, Chen M, Cao S, Zhang L, et al. Electrochemically assisted co-deposition of calcium phosphate/collagen coatings on carbon/carbon composites. Applied Surface Science. 2011;**257**:3612-3619

[62] Zhitomirsky I, Gal-Or L. Electrophoretic deposition of hydroxyapatite. Journal of Materials Science: Materials in Medicine. 1997;**8**:213-219

[63] Tahmasbi Rad A, Solati-Hashjin M, Osman NAA, Faghihi S. Improved bio-physical performance of hydroxyapatite coatings obtained by electrophoretic deposition at dynamic voltage. Ceramics International. 2014;**40**:12681-12691

[64] Sankar M, Suwas S, Balasubramanian S, Manivasagam G. Comparison of electrochemical behavior of hydroxyapatite coated onto WE43 Mg alloy by electrophoretic and pulsed laser deposition. Surface and Coatings Technology. 2017;**309**:840-848

[65] Peng P, Kumar S, Voelcker NH, Szili E, Smart RSC, Griesser HJ. Thin calcium phosphate coatings on titanium by electrochemical deposition in modified simulated body fluid. Journal of Biomedical Materials Research Part A. 2006;**76**:347-355

[66] Wang L-N, Luo J-L. Preparation of hydroxyapatite coating on CoCrMo implant using an effective electrochemically-assisted deposition pretreatment. Materials Characterization. 2011;**62**:1076-1086

[67] Kuo MC, Yen SK. The process of electrochemical deposited hydroxyapatite coatings on biomedical titanium at room temperature. Materials Science and Engineering: C. 2002;**20**:153-160

[68] Kannan S, Balamurugan A, Rajeswari S. Hydroxyapatite coatings on sulfuric acid treated type 316L SS and its electrochemical behaviour in Ringer's solution. Materials Letters. 2003;**57**:2382-2389

[69] Blanda G, Brucato V, Pavia FC, Greco S, Piazza S, Sunseri C, et al. Galvanic deposition and characterization of brushite/hydroxyapatite coatings on 316L stainless steel. Materials Science and Engineering: C. 2016;**64**:93-101

[70] Azem FA, Delice TK, Ungan G, Cakir A. Investigation of duty cycle effect on corrosion properties of electrodeposited calcium phosphate coatings. Materials Science and Engineering: C. 2016;**68**:681-686

[71] Kannan MB. 13—Hydroxyapatite coating on biodegradable magnesium. Hydroxyapatite (HAp) for Biomedical Applications; 2015. p. 289-306

[72] Popa MV, Moreno JMC, Popa M, Vasilescu E, Drob P, Vasilescu C, et al. Electrochemical deposition of bioactive coatings on Ti and Ti–6Al–4V surfaces. Surface and Coatings Technology. 2011;**205**:4776-4783

[73] Blackwood D, Seah K. Galvanostatic pulse deposition of hydroxyapatite for adhesion to titanium for biomedical purposes. Materials Science and Engineering: C. 2010;**30**:561-565

[74] Gopi D, Collins Arun Prakash V, Kavitha L, Kannan S, Bhalaji PR, Shinyjoy E, et al. A facile electrodeposition of hydroxyapatite onto borate passivated surgical grade stainless steel. Corrosion Science. 2011;**53**:2328-2334

[75] Sridhar TM, Kamachi Mudali U, Subbaiyan M. Preparation and characterisation of electrophoretically deposited hydroxyapatite coatings on type 316L stainless steel. Corrosion Science. 2003;**45**:237-252

[76] Isa NNC, Mohd Y, Yury N. Electrochemical deposition and characterization of hydroxyapatite (HAp) on titanium substrate. APCBEE Procedia. 2012;**3**:46-52

[77] Kar A, Raja K, Misra M. Electrodeposition of hydroxyapatite onto nanotubular TiO_2 for implant applications. Surface and Coatings Technology. 2006;**201**:3723-3731

[78] Cheng X, Wang J, Wang Y, Wang G, Zhao L. Preparation of thin hydroxyapatite layers on cp titanium through anodic oxidation followed with hydrothermal treatment. Sheng wu yi xue gong cheng xue za zhi= Journal of biomedical engineering= Shengwu yixue gongchengxue zazhi. 2002;**19**:378-382

[79] Le Guéhennec L, Soueidan A, Layrolle P, Amouriq Y. Surface treatments of titanium dental implants for rapid osseointegration. Dental Materials. 2007;**23**:844-854

[80] He L-p, Wu Z-j, Chen Z-z. In-situ growth of nanometric network calcium phosphate/porous Al_2O_3 biocomposite coating on Al Ti substrate. Chinese Journal of Nonferrous Metals. 2004;**14**:460-464

[81] Yang B, Uchida M, Kim H-M, Zhang X, Kokubo T. Preparation of bioactive titanium metal via anodic oxidation treatment. Biomaterials. 2004;**25**:1003-1010

[82] Zhang Y-y, Tao J, Pang Y-c, Wang W, Wang T. Electrochemical deposition of hydroxyapatite coatings on titanium. Transactions of Nonferrous Metals Society of China. 2006;**16**:633-637

[83] Kwok CT, Wong PK, Cheng FT, Man HC. Characterization and corrosion behavior of hydroxyapatite coatings on Ti6Al4V fabricated by electrophoretic deposition. Applied Surface Science. 2009;**255**:6736-6744

[84] Yang G-l, He F-m, Hu J-a, Wang X-x, Zhao S-f. Biomechanical comparison of biomimetically and electrochemically deposited hydroxyapatite–coated porous titanium implants. Journal of Oral and Maxillofacial Surgery. 2010;**68**:420-427

[85] Catauro M, Bollino F, Giovanardi R, Veronesi P. Modification of Ti6Al4V implant surfaces by biocompatible TiO_2/PCL hybrid layers prepared via sol-gel dip coating: Structural characterization, mechanical and corrosion behavior. Materials Science and Engineering: C. 2017;**74**:501-507

[86] Aktug SL, Kutbay I, and Usta M. Characterization and formation of bioactive hydroxyapatite coating on commercially pure zirconium by micro arc oxidation, Journal of Alloys and Compounds, 695, pp. 998-1004, 2/25/ 2017.

[87] Khoshsima S, Yilmaz B, Tezcaner A, Evis Z. Structural, mechanical and biological properties of hydroxyapatite-zirconia-lanthanum oxide composites. Ceramics International. 2016;**42**:15773-15779

[88] Pei X, Zeng Y, He R, Li Z, Tian L, Wang J, et al. Single-walled carbon nanotubes/hydroxyapatite coatings on titanium obtained by electrochemical deposition, Applied Surface Science. 2014;**295**:71-80

[89] Yuan Q, Golden TD. Electrochemical study of hydroxyapatite coatings on stainless steel substrates, Thin Solid Films. 2009;**518**:55-60

[90] Rojaee R, Fathi M, Raeissi K, Sharifnabi A. Biodegradation assessment of nanostructured fluoridated hydroxyapatite coatings on biomedical grade magnesium alloy. Ceramics International. 2014;**40**:15149-15158

[91] Janković A, Eraković S, Vukašinović-Sekulić M, Mišković-Stanković V, Park SJ, Rhee KY. Graphene-based antibacterial composite coatings electrodeposited on titanium for biomedical applications. Progress in Organic Coatings. 2015;**83**:1-10

[92] Albayrak O, El-Atwani O, Altintas S. Hydroxyapatite coating on titanium substrate by electrophoretic deposition method: Effects of titanium dioxide inner layer on adhesion strength and hydroxyapatite decomposition, Surface and Coatings Technology. 2008;**202**:2482-2487

[93] Li H, Khor K, Cheang P. Titanium dioxide reinforced hydroxyapatite coatings deposited by high velocity oxy-fuel (HVOF) spray. Biomaterials. 2002;**23**:85-91

[94] Vassen R, Stuke A, Stöver D. Recent developments in the field of thermal barrier coatings. Journal of Thermal Spray Technology. 2009;**18**:181-186

[95] Song B, Pala Z, Voisey KT, Hussain T. Gas and liquid-fuelled HVOF spraying of Ni50Cr coating: Microstructure and high temperature oxidation. Surface and Coatings Technology. 2016;**318**:224-232

[96] Bolelli G, Rauch J, Cannillo V, Killinger A, Lusvarghi L, Gadow R. Microstructural and tribological investigation of high-velocity suspension flame sprayed (HVSFS) Al_2O_3 coatings. Journal of Thermal Spray Technology. 2009;**18**:35-49

[97] Bolelli G, Giovanardi R, Lusvarghi L, Manfredini T. Corrosion resistance of HVOF-sprayed coatings for hard chrome replacement. Corrosion Science. 2006;**48**:3375-3397

[98] Zhang S-L, Li C-X, Li C-J, Yang G-J, Liu M. Application of high velocity oxygen fuel flame (HVOF) spraying to fabrication of La0.8Sr0.2Ga0.8Mg0.2O3 electrolyte for solid oxide fuel cells. Journal of Power Sources. 2016;**301**:62-71

[99] Killinger A, Kuhn M, Gadow R. High-Velocity Suspension Flame Spraying (HVSFS), a new approach for spraying nanoparticles with hypersonic speed. Surface and Coatings Technology. 2006;**201**:1922-1929

[100] Ban Z-G, Shaw L. Characterization of thermal sprayed nanostructured WC-Co coatings derived from nanocrystalline WC-18wt.% Co powders. Journal of Thermal Spray Technology. 2003;**12**:112-119

[101] Visai L, De Nardo L, Punta C, Melone L, Cigada A, Imbriani M, et al. Titanium oxide antibacterial surfaces in biomedical devices. International Journal of Artificial Organs. 2011;**34**:929-946

[102] Dongmo E, Wenzelburger M, Gadow R. Analysis and optimization of the HVOF process by combined experimental and numerical approaches. Surface and Coatings Technology. 2008;**202**:4470-4478

[103] Pan J, Hu S, Niu A, Ding K, Yang L. Numerical analysis of particle impacting and bonding processes during high velocity oxygen fuel spraying process. Applied Surface Science. 2016;**366**:187-192

[104] Singh H, Kaur M, Bala N. High velocity oxy-fuel spraying and surface finish. In: Reference Module in Materials Science and Materials Engineering. Comprehensive Materials Finishing. Elsevier. 2016;**3**:207-219

[105] Rauch J, Bolelli G, Killinger A, Gadow R, Cannillo V, Lusvarghi L. Advances in High Velocity Suspension Flame Spraying (HVSFS). Surface and Coatings Technology. 2009;**203**:2131-2138

[106] Gadow R, Killinger A, Stiegler N. Hydroxyapatite coatings for biomedical applications deposited by different thermal spray techniques. Surface and Coatings Technology. 2010;**205**:1157-1164

[107] Gross KA, Berndt CC. Biomedical application of apatites. Reviews in Mineralogy and Geochemistry. 2002;**48**:631-672

[108] Gross K, Chai C, Kannangara G, Ben-Nissan B, Hanley L. Thin hydroxyapatite coatings via sol–gel synthesis. Journal of Materials Science: Materials in Medicine. 1998;**9**:839-843

[109] Bolelli G, Bellucci D, Cannillo V, Gadow R, Killinger A, Lusvarghi L, et al. Comparison between Suspension Plasma Sprayed and High Velocity Suspension Flame Sprayed bioactive coatings. Surface and Coatings Technology. 2015;**280**:232-249

[110] Bolelli G, Bellucci D, Cannillo V, Lusvarghi L, Sola A, Stiegler N, et al. Suspension thermal spraying of hydroxyapatite: Microstructure and in vitro behaviour. Materials Science and Engineering: C. 2014;**34**:287-303

[111] Levingstone TJ, Ardhaoui M, Benyounis K, Looney L, Stokes JT. Plasma sprayed hydroxyapatite coatings: Understanding process relationships using design of experiment analysis. Surface and Coatings Technology. 2015;**283**:29-36

[112] Sansongsiri S, Kaewmanee T, Boonyawan D, Yu LD, Thongtem S. Effect of titanium dioxide blocking layer deposited by cathodic arc plasma on the energy conversion efficiency of dye-sensitized solar cells. Surface and Coatings Technology. 2016;**306**(Part A): 257-261

[113] Shahien M, Suzuki M. Low power consumption suspension plasma spray system for ceramic coating deposition. Surface and Coatings Technology. 2017;**318**:11-17

[114] Vautherin B, Planche MP, Montavon G, Lapostolle F, Quet A, Bianchi L. Study of metallic powder behavior in very low pressure plasma spraying (VLPPS)—Application to the manufacturing of titanium–aluminum coatings. Surface and Coatings Technology. 2015;**275**:341-348

[115] Morks M. Fabrication and characterization of plasma-sprayed HA/SiO_2 coatings for biomedical application. Journal of the Mechanical Behavior of Biomedical Materials. 2008;**1**:105-111

Chapter 6

Biomimetic Calcium Phosphates Derived from Marine and Land Bioresources

Florin Miculescu, Aura-Cătălina Mocanu,
Andreea Maidaniuc, Cătălina-Andreea Dascălu,
Marian Miculescu, Ștefan Ioan Voicu and
Robert-Cătălin Ciocoiu

Additional information is available at the end of the chapter

http://dx.doi.org/10.5772/intechopen.71489

Abstract

This chapter aims to establish the key factors for technological optimization of biogenic calcium phosphate synthesis from marine and land resources. Three natural calcium sources—marble, seashell and bovine bone—were considered as raw materials. The proposed materials are suitable candidates for the synthesis of bone substitutes similar to the inorganic bone component. The synthesis processes were developed based on the investigations of thermal phenomena (TGA-DSC analysis) that can occur during thermal treatments. By this method, we were able to determine the optimum routes and temperatures for the complete dissociation of calcium carbonate as well as risk-free deproteinization of bovine bone. An exhaustive characterization, performed with modern and complementary techniques such as morphology (SEM), composition (EDS, XRF) and structure (FT-IR, XRD), is presented for each precursor. The final chemical composition of ceramic products can be modulated through a careful control of the key parameters involved in the conversion, in order to create long-term performant biphasic apatite biomaterials, with broad medical applicability. Identifying the suitable strategies for this modulation contributes to an appreciable advance in orthopedic tissue engineering.

Keywords: phosphate biomaterial synthesis and processing, marine and land bioresources, biomimetic calcium phosphates, modulated calcium carbonate-derived HA proportion, bovine bone-derived HA

1. Introduction

The fifth of twelve principles of Green Chemistry states that: *"The use of auxiliary substances (e.g., solvents, separation agents, etc.) should be made unnecessary whenever possible and innocuous when used"* [1].

During the last years, the increased awareness of the scientific community regarding clean preparation and processing of bulk and powder biomaterials resulted in intense use of alternative precursors for preparing adequate materials for orthopedic medical applications. However, the development of a bone reconstruction material, completely biologically and mechanically compatible with the different types of bone tissue, is still an ongoing challenge.

Human bone anatomy has the architecture of a nanocomposite material, made of 60–70% mineral component, up to 30% organic components (mostly type I collagen) and approximately 10% water. The mineral component, usually defined as biological apatite and sometimes misrepresented as natural hydroxyapatite, incorporates multiple substitutions [2, 3]:

- Calcium (Ca^{2+}) can be substituted by Sr^{2+}, Ba^{2+}, Mg^{2+}, Na^{+} or K^{+};
- Phosphorus (P) can be substituted by C, As, V or S; and
- Hydroxyl groups (OH^{-}) can be substituted by carbonate groups $(CO_3)^{2-}$, fluorine (F^{-}), chlorine (Cl^{-}) or their place can remain vacant.

The need for restoring damaged bone tissues leads to the development of various bone reconstruction and tissue engineering solutions. Currently, the most popular are various types of bone grafts but every type is confronted to disadvantages such as the risk of biological contamination, infection and fast absorption (for xenografts); difficult harvesting and storage, high risk of tumoral cells and pathogens transfer (for allografts); low availability, additional surgical procedures, scars and prolonged healing of harvested area (for autografts) [4, 5].

Alloplastic materials are intensively developed as alternatives for bone grafts. The current market offers a wide range of calcium phosphate-based biomaterials as substitutes for bone tissue. Most representative materials are hydroxyapatite (HA), beta-tricalcium phosphate (β-TCP) and different combinations of these, generally named biphasic calcium phosphates (BCP) [2, 3, 6–8]. The main reason for using calcium phosphate-based biomaterials is their resemblance with the bone tissue, so research and development of this area tends to reproduce more accurately the damaged tissue, with more efficient results. This scope involves firstly the preparation of a calcium phosphate with potential use in orthopedic bone reconstruction.

Currently, this trend is expressed by improving a relatively new concept, which combines advanced fabrication of bioceramics with the sustainable use of natural resources, namely functionalization of marine and land resources for preparing biogenic calcium phosphates [2, 9, 10]. Dedicated studies offer extensive information regarding [1] *marine resources* such as vertebrates bones—fish bones [5, 6, 11, 12] and calcified structures of invertebrates—coral, snail, seashell, cuttlefish, sea urchin, etc. [12–17], and [2] *land resources* such as animal bone tissue—preponderantly bovine bone [2, 4, 18, 19] or other calcified structures such as eggshells [9, 20], which could be used as cost-effective raw materials. Most of these resources are naturally available as various polymorphs of calcium carbonate ($CaCO_3$); by exception, vertebrate bones contain calcium phosphate closely related to the mineral component of human bones. Another $CaCO_3$ resource available on land is marble. This is, to our knowledge, an innovative precursor for preparing biocompatible calcium phosphates.

All these natural resources bring, besides the calcium resources required for calcium phosphate synthesis, many beneficial chemical elements such as Mg, Na, K, Cl, F, Si, Sr, which are found in human bones in various amounts [2]. This provides a unique advantage to naturally derived biomaterials against conventional calcium phosphates because the synthesis and preparation methods do not need additional procedures for doping the stoichiometric compounds.

Various synthesis methods were developed so far for using $CaCO_3$ as a precursor for obtaining calcium phosphates, among which, two research directions being currently considered:

- *Direct synthesis*, which implies precursor treatment with phosphorus-based reagents. Although intensively studied [6, 14, 15, 21–24], the process parameters are incompatible with reproducible manufacturing, while the final products are susceptible to impurification with trapped intermediate products.
- *Indirect synthesis*, which requires thermal dissociation of $CaCO_3$ in calcium oxide (CaO) prior to phosphate-based reagent treatment. Initially proposed by Rathje in 1939 [25], the method was studied for different types of precursors in an attempt to obtain reproducible results [5, 6, 11, 13, 16–18, 26]. To date, the correlation between the synthesis parameters and the material characteristics is poorly understood, so further research is needed for adapting the method for advanced manufacturing.

For animal bones, isolation and processing of the existent calcium phosphate resource begins with chemical or thermal deproteinization; then, the resulted material is thermally treated in a controlled manner: thermal treatment could be performed by combining different temperatures (700–1400°C), heating environments (air, argon, nitrogen, carbon dioxide) and cooling conditions (air or water with ice) [27, 28]. Although methods for manufacturing bovine bone-derived medical devices are standardized [29] and their use is regulated [5], the current research strategies aim to align bovine bone processing to the fifth principle of *Green Chemistry* and to improve the existent methods by eliminating all reagents which could induce a risk, thus upgrading the quality management approaches related to bovine bone-derived products, their manufacturing and large-scale utilization.

This study aims to identify the key parameters for optimization of biogenic calcium phosphate synthesis and processing. The marine and land resources included in this study are marble, seashells and bovine bone. The proposed natural resources are convenient candidates for preparing bone substitutes, which resemble the inorganic (mineral) component of natural bone tissue. Moreover, by careful control of key parameters involved in $CaCO_3$ conversion and biological apatite isolation, the composition of final ceramic products could be modulated in order to create long-term performant biphasic calcium phosphates with larger biomedical applicability. Identifying the optimal routes for achieving this aims contributed to a substantial advancement of bone reconstruction materials.

2. Marine and land bioresources

The worldwide scientific community is aware of the negative environmental effects of human consumption. A continuous effort aims to reduce the impact of unsustainable use of limited resources by developing environment-friendly processing methods and applications [9]. The use of marine and land materials resulted after industrial processing of different animal species for producing performant biomaterials is a sustainable solution for reducing waste generation.

2.1. Marine resources

Different invertebrate organisms from marine environment contain considerable resources of $CaCO_3$ in different calcified structures such as thorns, shells, exoskeletons or bones. Some of the most popular marine precursors used for the preparation of hydroxyapatite and other calcium phosphates were corals, due to a well-established conversion procedure for coralline hydroxyapatite which was developed by Roy and Linnehan in 1974 [22]. However, current threats such as climate change, destructive fishing practices, overfishing, careless tourism, pollution or coral mining (for use as bricks, road-fill, cement or souvenirs) drastically limited the possibility of using these resources for producing hydroxyapatite. In this respect, different available alternatives can be used:

- Bone-like structures in cuttlefish (e.g., *Sepia officinalis*), generally known as cuttlebones, are organized in $CaCO_3$ (aragonite) pillars and organic membranes (3–4.5% organic matter) [30]. Synthesis of hydroxyapatite is usually achieved by direct synthesis with phosphorus-based reagents.
- Gastropods (snails) and bivalves (clams, mussels, oysters) shells are primarily composed of $CaCO_3$. Many types of shells (e.g., *Strombus gigas, Tridacna gigas, Mytilus edulis*) are considered lamellar microcomposites. The inner layers of these shells consist of 95–99% of $CaCO_3$ as aragonite and different organic macromolecules [13]. Based on their environment, the different species of shells may contain variable amounts of oxides such as SiO_2, MgO, Al_2O_3, SrO, P_2O_5, Na_2O or SO_3 [14–16].
- Echinoderm skeletons (e.g., *Heterocentrotus mamillatus, Heterocentrotus trigonarius*) are composed of $CaCO_3$ plates and spines, each skeletal element being a single crystal of $CaCO_3$ in form of calcite, very finely branched and structured, for which conversion in hydroxyapatite was reported [14].
- Crustacean wastes (crabs or lobsters such as *Portunus pelagicus, Nephrops norvegicus*, etc.) contain three types of valuable compounds—20–40% protein, 20–50% $CaCO_3$ and 15–40% protein. Current waste processing is destructive, wasteful and expensive, as the methodology required to separate these three types of compounds uses corrosive or hazardous reagents. Creative chemistry is summoned in order to fully benefit from this type of waste, which is largely available [31].

Besides $CaCO_3$, calcium phosphate sources are available in fish bones:

- Fish bones (e.g., *Thunnus obesus, Pseudoplatystoma corruscans, Pseudoplatystoma fasciatum, Oreochromis mossambicus, Paulicea lutkeni*, etc.) represent a significant part of the fish—10–15% of total fish biomass being bones from the head to vertebrae. Although interspecies variation of composition of fish bones is significant in the level of proteins and lipids, the mineral bone matrix contains similar amounts of Ca and P, giving a similar Ca/P ratio regardless of the species.

2.2. Land bioresources

Similar to marine bioresources, terrestrial or land raw materials can be divided into $CaCO_3$ and calcium phosphate resources:

- Land gastropod shells (snails such as *Helix pomatia* or *Helix aspersa*) contain mainly $CaCO_3$ and minor amounts of $MgCO_3$ and organic compounds, which were reported as raw materials for producing natural bioceramics [32, 33].
- Land crustaceans (e.g., *Orchestia cavimana*) contain amorphous $CaCO_3$ and minor amounts of amorphous calcium phosphate within their structures.
- Bird eggshells (*Gallus gallus domesticus, Struthio camelus*) contain up to 97% $CaCO_3$ as calcite and 3–4% organic components. Conversion of $CaCO_3$ in biocompatible calcium phosphates can be achieved by direct synthesis with phosphate-based reagents [9, 20].
- Large vertebrate bones are primarily composed of calcium phosphate (biological apatite) and are largely available worldwide. Procedures for isolating the mineral component of several bone species (*Cervidae, Ovis aries, Equus caballus, Crocodylinae, Struthio camelus, Anatidae*) were already reported. Use of bovine bones is considered a more practical approach in terms of size, availability and similarity with human bones [2, 4, 5]. The biological apatite of bovine bones also includes $Mg2^+$ ions and CO_3^{2-} groups which further influence the characteristics of processed materials [2].

3. Synthesis and preparation

3.1. Precursor's preparation: impurities and organic components removal

Independent of the resource used for calcium phosphate preparation, the raw material shall be subjected to preliminary preparation procedures in order to ensure the quality of the final products. Generally, these procedures refer to the macroscopical impurities removal and organic components separation from the natural material's structure.

For invertebrates, cleaning of the precursors can be accomplished by brushing under water pressure and distilled water ultrasonication [15, 17]. The residual organic matter can be removed by immersion in hydrogen peroxide solution (50%), through boiling or in autoclave [13]. After drying, materials can be crushed and grounded in a ball mill or agate mortar and optionally sieved [15, 17].

Fish bones can be first mechanically cleaned to remove impurities/particles from the natural environment and then sectioned into small pieces [4, 12]. Further, the bone can be repeatedly boiled in distilled water to separate the organic tissue and bone marrow [4, 6, 11]. Degreasing and elimination of external hyaluronic acid and proteins can be achieved by bones immersion either in alcohol baths (ethanol 70%, v/v), followed by distilled water washing and hydrogen peroxide preservation (30%, v/v) or in alkaline sodium hydroxide (NaOH) solution (1 N) [4, 12]. After drying at 50°C in hot air oven, the bone pieces can be stored in formaldehyde solution (4%, v/v) if it is not immediately processed [4].

Bird or tetrapod bone preparation starts by freezing at −20°C for facile segmentation; otherwise, mechanical removal of macroscopical impurities may lead to local heating of the bone and therefore to inadequate separation of bone marrow and other potential antigenic substances [2].

Further deproteinization can be carried out in an autoclave at 100°C by boiling [2, 19]. Prolonged exposure to autoclaves or vapors reduces collagen in the form of gelatin and thus lowers the risk of coal black matter appearance at the end of processing [2]. This step can also be achieved with organic solvents such as methyl acetate or hydrogen peroxide [2].

For eggshells, cleaning with sodium hypochlorite (NaClO) solution (5%) was reported, followed by ball mill or agate mortar grounding for 2 hours and sieving. Further, the obtained powder was repeatedly washed with the same solution and then dried in conventional oven at 100°C for 24 hours [9, 20].

3.2. Synthesis by chemical precipitation

3.2.1. Direct method

Direct synthesis (**Figure 1**) is performed on $CaCO_3$ powder with phosphorus-based reagents. One commonly studied reagent is ammonium phosphate monohydrate ($(NH_4)_2HPO_4$), which

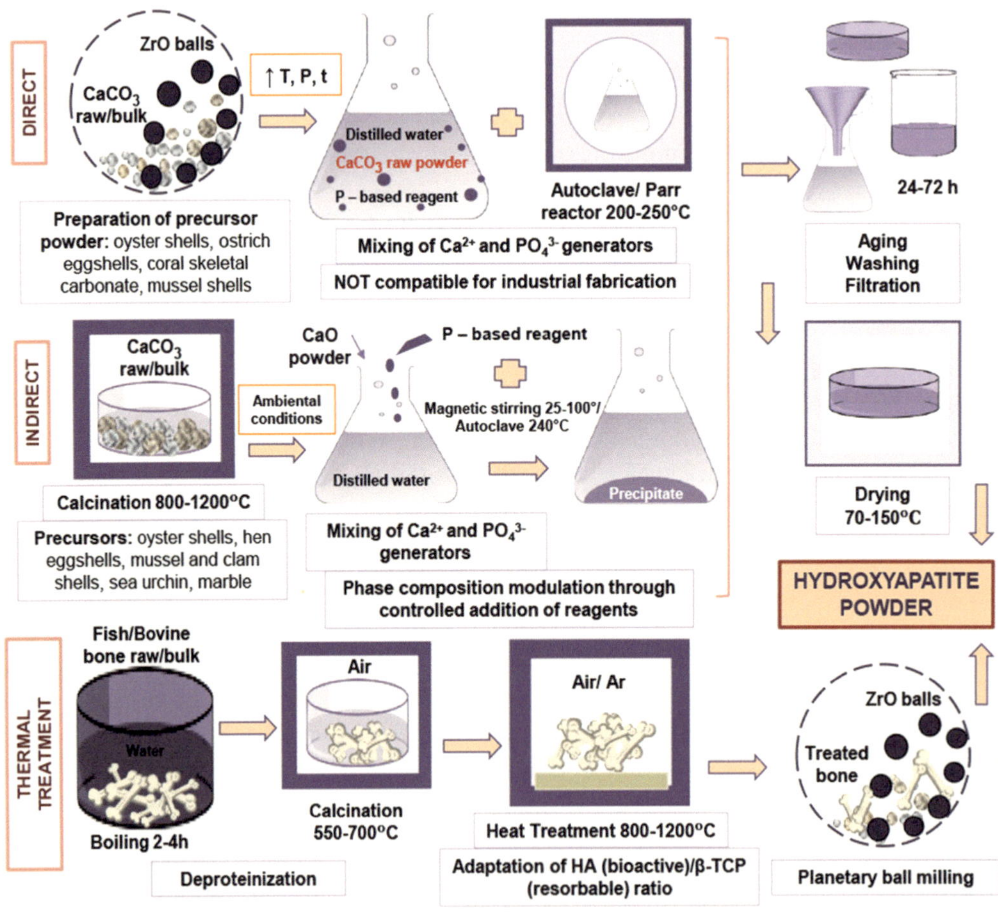

Figure 1. Schematic representation of the main routes for converting natural precursors such as marble, seashell, and bovine bone in biocompatible calcium phosphates.

can be mixed with $CaCO_3$ powder in Parr reactors at approximately 250°C [22] or in an autoclave. $CaCO_3$ powder can also be mixed with distilled water and then treated by controlled addition with phosphoric acid (H_3PO_4) in equivalent proportions for desired Ca/P molar ratio [20]. Another possible method concerns the direct treatment with calcium pyrophosphate ($Ca_2O_7P_2$) by wet grounding in a planetary ball mill [15]. In all cases, the synthesis is followed by distilled water washing and drying at temperatures between 70 and 150°C [15, 20, 22–24].

3.2.2. Indirect method

3.2.2.1. Thermal dissociation of calcium carbonate

Regardless of the $CaCO_3$ polymorphic form (aragonite, calcite or dolomite), thermal dissociation takes place through calcination at temperature of 800–1200°C, for at least 2 hours [5, 6, 16, 17, 34]. Calcination leads to carbon dioxide (CO_2) release, associated with a mass loss of ~45%. The obtained product is calcium oxide powder (CaO), which can be involved in the chemical synthesis as prepared or sieved in advance [16, 34].

3.2.2.2. Chemical precipitation

Post calcination, CaO powder is usually mixed with distilled water and transformed into calcium hydroxide ($Ca(OH)_2$); during hydration, an exothermic reaction occurs and its volume doubles [34]. After hydration, the $Ca(OH)_2$ aqueous solution can be further treated with phosphor-based reagents such as diammonium phosphate ($(NH_4)_2HPO_4$) [5, 13] or H_3PO_4. CaO powder can also be dissolved in nitric acid (HNO_3), to obtain calcium nitrate ($Ca(NO_3)_2$) which would then react with $(NH_4)_2HPO_4$ [17] or EDTA solution, to convert CaO into Ca-EDTA complex, which reacts with disodium phosphate (Na_2HPO_4) [16].

Reactions take place for several hours by using magnetic stirring at temperature of 25–100°C [5, 16, 26] or several days, in autoclave, at maximum 240°C [13]. The reaction ends with formation of a white precipitate which can be further dried in a vacuum oven at 80°C for 6 hours [16] or in an electric one for 3 hours [5, 17, 35] to obtain the final calcium phosphate powder. Following this procedure, after sintering at 900°C, the obtained HA was reported to be pure and thermally stable.

Depending on the medical applicability, the product's final composition can be tuned during synthesis and through the final thermal treatment. This is the main reason why elevated synthesis conditions (direct method) are not necessarily adequate for synthesizing large amounts of powdery samples. Therefore, in case of calcium carbonate-based precursors, synthesis techniques were adapted to normal (room) conditions such as temperature, pressure and time (indirect method). On the other hand, for bone-like precursors, the synthesis route is completely different, given their compositional similarity to the human bone. The HA extraction and final composition adaptation is carried out only by thermal treatment.

3.3. Thermal treatment

The thermal treatment of marine resources can be performed at 160–1200°C for 2–8 hours [6, 12, 21]. A soaking time of 2 hours proved insufficient for complete transformation of

$CaCO_3$, and small quantities of residual aragonite could be identified in the material structure. Extension of thermal treatment to 8 hours was reported to ensure the complete conversion of calcium carbonate into HA [14]. Even though it was reported that at 1000°C synthesized HA is stable and similar to the pure one with Ca/P molar ratio of 1.67 [21], HA preparation from marine resources and exclusively thermal methods is not reproducible.

On the other hand, thermal treatment of bone tissue aims for producing biphasic calcium phosphates with modulated content of HA and β-TCP, since β-TCP transformation into α-TCP was not identified in bovine bone-derived materials at temperatures lower than 1200°C [27, 36]. Adaptation of HA (bioactive)/β-TCP (resorbable) ratio relies upon the precursor features such as substitutions in the crystalline structure of biological apatite, the elemental species embedded in the structure and the interactions between them during thermal processing. Apart from the biological apatite characteristics, an important role for processing is addressed to the thermal treatment parameters:

3.3.1. *Heating rate and heating duration*

Temperature and heating duration are dependent on the bone pieces' dimensions, the amount of oxygen present in the heating environment and precursor preparation methods [37]. At 600–1000°C, at least 2 hours is necessary for the removal of all organic component from 1 cm^3 of bone tissue. Optimum reported heating rate was 10°C/min; thermal treatments conducted below this rate (5°C/min) could lead to the partial fixation of carbon and delayed decomposition reactions [28].

3.3.2. *Treatment temperature*

Thermal degradation of bovine bone begins with the evaporation of surface water. Collagen denaturation is carried out in parallel with the water loss and continues up to 500–600°C, with mass losses and carbon dioxide emissions [28]. Until complete degradation, the organic component acts as a protective shield for calcium phosphate found in the bone mineral component. For this reason, the mineral matter does not undergo thermal transformations up to 500–600°C. Above this temperature, the biological apatite is subjected to a recrystallization process, made in three stages: lattice diffusion (500–750°C), surface diffusion (750–900°C) and grain boundary diffusion (900–1000°C) [38]. Recrystallization is usually correlated with removal of carbonate groups from the crystalline structure. Thermal degradation of bone-derived HA is possible above 1000°C but the event is strongly influenced by the precursor's chemical composition (with compositional variations of bone tissue from different animals) and the thermal treatment environment. The main products obtained after HA decomposition include different forms of oxyapatites, which can subsequently decompose into β-TCP, CaO [36] or tetracalcium phosphate (TTCP).

3.3.3. *Thermal treatment environment*

Thermal treatment environment is responsible for the heat transfer and assuring/disposing of gaseous products and reactants. Thermal analyses performed in nitrogen atmosphere proved that bovine bone-derived HA decomposition begins at approximately 1000°C and lead to the β-TCP and CaO traces formation. Other result obtained in air atmosphere and from different species of vertebrates pointed the beginning of HA transformation at around 800°C [39]. In argon atmosphere, bovine bone-derived HA decomposed into β-TCP (without any detectable CaO traces) at temperature of ~1200°C [36]. Heating in carbon dioxide atmosphere does not induce significant modifications of HA up to 1200°C.

3.3.4. Cooling conditions

The control of cooling conditions contributes to the modification/preservation of the bone-derived calcium phosphates' phase composition because the conversion β-TCP → α-TCP is reversible through slow cooling. α-TCP (resorbable) conservation within the thermally processed calcium phosphates' structure was achieved by quenching.

4. Results and discussion

The investigations described in this section were made on samples prepared from natural precursors—marble, marine seashells and bovine bone—and from bioceramic materials derived from those precursors. For marble and seashells, an indirect Rathje-based method was optimized by means of magnetic stirring, reagent treatment and thermal treatment [40, 41]. In this study, the materials were evaluated in three stages: [1] raw marble and seashell precursors (after cleaning), [2] thermally treated marble and seashells (intermediate products), and [3] marble and seashell-derived bioceramics (named Marble-TT and Seashell-TT, respectively) resulted after treatment with H_3PO_4 and drying at 120°C.

Bovine bone samples were investigated in three stages: [1] raw bovine bone precursor; [2] intermediate product obtained after deproteinization at 500°C; and [3] final ceramic product obtained after thermal treatment at 1000°C in air atmosphere. Following this route, contamination risks were reduced by eliminating all reagents involved in processing and by performing thermal treatment at temperatures above 850°C, which are considered microbiologically safe [27, 28]. The preparation procedures for all precursor types are described in detail in Refs. [36, 40].

4.1. TGA/DSC analysis: thermal behavior of natural resources

Temperature-induced thermal transformations were evaluated by TGA-DSC analysis (SDT Q600 equipment) between 25 and 1200°C, with 10°C/min, in argon atmosphere. The results are presented in **Figure 2**.

Thermal degradation of calcium carbonates (marble and seashells) began with a thermal event associated with evaporation of surface water, at approximately 75°C, without significant mass loss (**Figure 2**). $CaCO_3$ thermal degradation included the decomposition of dolomite from marble and aragonite from seashells into calcite, a thermally stable phase. Decompositions occurred at approximately 300°C (endothermic peaks in **Figure 2**) and were accompanied by a mass loss of approximately 2%. Thermal dissociation of $CaCO_3$ continued until 850–900°C temperature was achieved. This dissociation temperature is inferior to the one corresponding to pure $CaCO_3$ (963°C). Degradation was associated with mass losses of approximately 40%, corresponding to CO_2 emissions. The CaO resulted after carbonate degradation was stable until 1200°C was achieved, in agreement with previous studies [42].

Thermal degradation of bovine bone began with surface water removal, which occurred until approximately 300°C and was accompanied by a mass loss of approximately 10% (**Figure 2**). Combustion of the organic bone component began concomitantly with water loss. This phenomenon was accompanied by a significant mass loss of 20–30%, until 500°C was reached. The temperatures between 500 and 800°C usually induce the removal of carbonate groups within materials' structure; in the current study, this event was associated with a mass loss of approximately 5%. First major thermal event was identified in DSC results at approximately 800°C and corresponds to the partial transformation of HA in β-TCP, between 850 and 1200°C, with an exothermic peak at approximately 1000°C. The beginning of a new thermal event was observed at approximately 1200°C, which suggests that β-TCP was partially transformed into α-TCP. This result is in agreement with the previously reported results, which pinpoint the beginning of α-TCP at 1125°C [27, 28].

4.2. SEM-EDS analysis: morphocompositional characteristics

The morphocompositional characteristics of marble, seashell and bovine bone precursors and bioceramic products derived from those precursors were highlighted by SEM analysis (Philips Xl 30 ESEM TMP equipment) coupled with EDS (EDAX Sapphire equipment). The

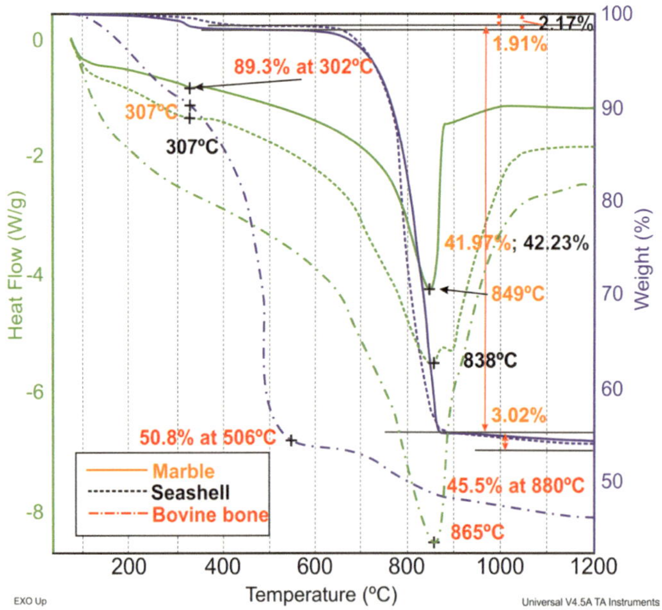

Figure 2. TGA-DSC analysis results for marble, seashell, and bovine bone precursors used for biocompatible calcium phosphate preparation.

results for precursors, intermediate synthesis products and final bioceramics derived from each type of natural resources are presented comparatively in **Figure 3**.

The raw marble (**Figure 3A**) exhibited a compact microstructure, with a separated phase arrangement. Considering the EDS results (in which magnesium presence was confirmed, shown in **Figure 3**), as well as the previously reported results, the fine white lines in the marble microstructure represent calcite microregions. This alternates with the broader regions of dolomite (magnesium and calcium carbonate), highlighted by darker gray shades. Isolated

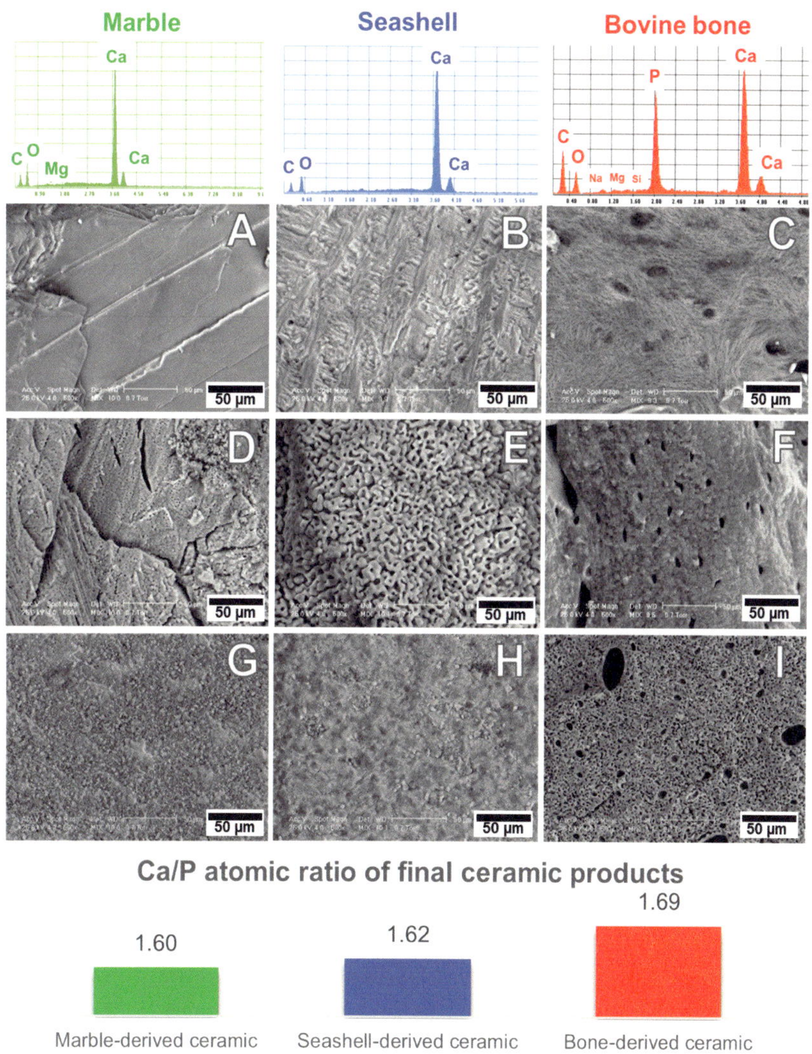

Figure 3. SEM-EDS results for marble, seashell, and bovine bone precursors and bioceramic products prepared from those precursors: EDS spectra of natural precursors; (A–C) SEM results for initial (raw) precursors; (D–F) SEM results for intermediate products: marble and seashell thermally treated at 1300°C and bovine bone thermally treated at 500°C. (G–I) SEM results for final bioceramic products (obtained after chemical treatment and drying of marble and seashell, and thermal treatment of bovine bone, respectively). Variation of Ca/P atomic ratio of final ceramic products.

calcite grains, with poliedric shape and sharp edges were also observed. After the thermal treatment performed at 1200°C (**Figure 3D**), the material exhibits an acute cracking of the initial compacted microstructure. The final bioceramic product (**Figure 3G**), obtained after chemical treatment with H_3PO_4 and drying at 120°C exhibits a dense and uniform microstructure with no pores or defects.

The seashell precursor (**Figure 3B**) had a typical lamello-fibrillar microstructure, in which calcite layers alternated with perpendicular aragonite layers. After thermal treatment (**Figure 3E**), shells morphology was constituted from connected particles and many pores resulted after cracking and aeration of calcite layers. After synthesis and final thermal treatment (**Figure 3H**), the resulted bioceramic had a compact, uniform and defect-free microstructure.

The raw bovine bone microstructure (**Figure 3C**) is typical for cortical bone tissue, that is, an association of osteons with concentric lamellae arranged around haversian canals. The organic component of bone tissue (highlighted by darker gray shades in **Figure 3C**) was mostly present in the haversian canals and in the lacunae disposed along bone lamellae. In thermal treated bone (at 500°C—**Figure 3F** and at 1000°C—**Figure 3I**), both haversian canals and lacunae were transformed into different sized pores due to complete combustion of the bone organic component.

The composition of carbonate precursors includes Ca, C and O as major elements (EDS spectra in **Figure 3**). Traces of Mg were identified in the EDS spectra of the marble precursor. Based on their origin, these precursors may contain variable quantities of Na and/or Si, but were not identified in the current study. The EDS spectra of the bovine bone precursor includes, besides Ca, P, C and O (characteristic major elements), peaks of Na and Mg.

The compositional key performance indicator for the naturally derived bioceramic was atomic Ca/P ratio (graph in **Figure 3**), calculated based on EDS results. Ca/P ratio varied between 1.60 for marble-derived materials and 1.69 for bovine bone-derived ones.

4.3. XRD: structure and phase composition

The structure and phase composition of natural precursors and final ceramic products were evaluated by XRD (Bruker D8 Advance diffractometer equipped with a LynxEye detector), in Bragg-Brentano geometry, with Cu K_α ($\lambda = 1.5418$ Å). Analyses were performed for $2\theta = 10–50°$, with 0.04°/1 s step. The results are presented in **Figures 4** and **5**.

The XRD patterns for marble and seashell precursors (**Figure 4**) indicated the presence of $CaCO_3$ by its characteristic peaks at ~29.5°, 47–48° (marble—calcite, ICDD: 01-086-2339) and ~27°, 32.5°, 43° (seashells—aragonite, ICDD: 00-005-0453), respectively. In agreement with EDS results (**Figure 3**), XRD pattern for marble (**Figure 4**) signaled the presence of magnesium carbonate $CaMg(CO_3)_2$ (ICDD: 00-036-0426) in the material, by characteristic peaks located at 37° and 42.5°. XRD results for the bovine bone precursor indicate a low crystallinity due to the presence of the organic components within the bone tissue.

The peaks identified in the XRD pattern of the bioceramic obtained after bone thermal treatment (bovine bone-derived ceramic in **Figure 5**) confirm that the materials contain HA as single phase. The sharpness of the peaks suggests a high crystallinity. In comparison, the bioceramics derived from marble and seashells (marble-derived ceramic and seashell-derived

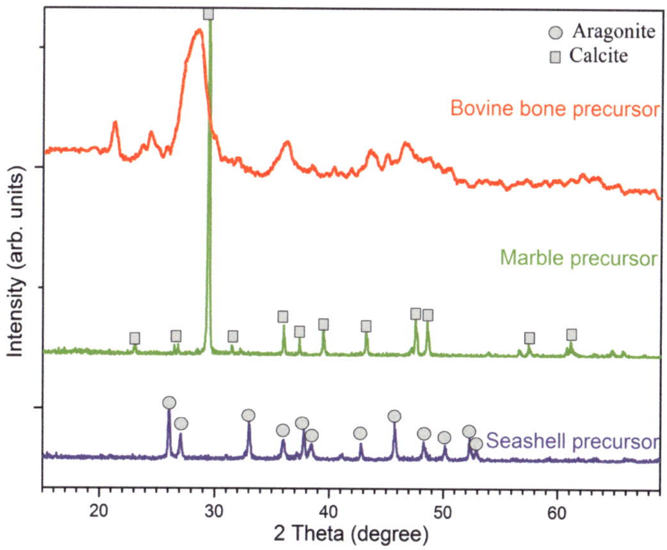

Figure 4. XRD patterns of marble, seashell, and bovine bone precursors used for biocompatible calcium phosphate preparation.

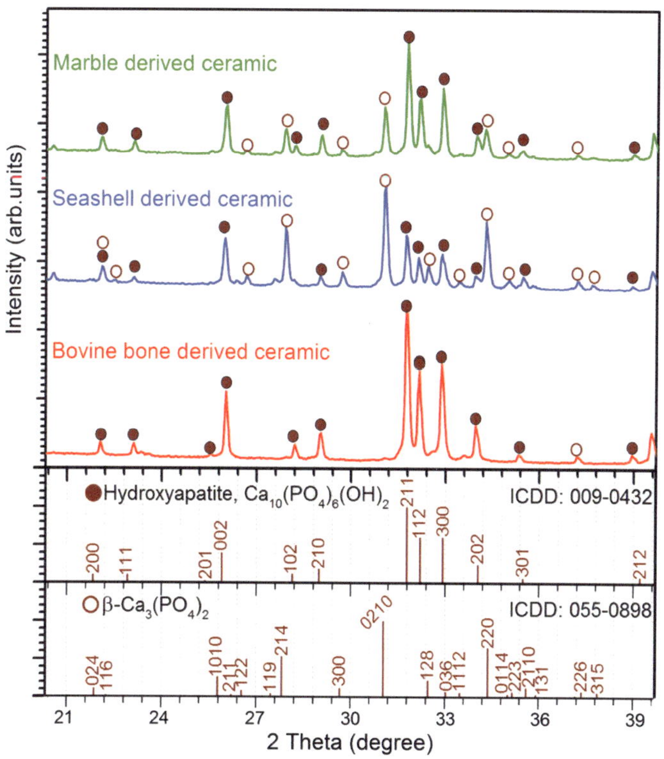

Figure 5. XRD patterns of final derived bioceramic products obtained after indirect chemical synthesis and thermal treatment of marble and seashell precursors (marble-derived ceramic and seashell-derived ceramic, respectively) and thermal treatment at 1000°C of cortical bovine bone (bovine bone-derived ceramic).

ceramic in **Figure 5**) included low-intensity peaks suggesting a lower crystallinity. The indirect synthesis of both precursor types led to the obtaining of a biphasic material consisting of different proportions of HA and β-TCP.

4.4. FT-IR analysis: functional groups architecture

Functional groups architecture was evaluated by FT-IR analysis (Perkin Elmer Spectrum BX II equipment) in attenuated total reflectance (ATR) mode (PikeMiracle head). IR spectra were recorded between 800 and 3600 cm^{-1} for the raw precursors (**Figure 6**) and between 500 and 1200 cm^{-1} (**Figure 7**) for the final bioceramic products, with 4 cm^{-1} resolution and 32 scans per experiment.

IR spectra of the marble and shell precursors (**Figure 6**) included the characteristic vibration bands of CO_3^{2} groups in $CaCO_3$, namely ν_2 asymmetric bending (870 cm^{-1}), ν_3 asymmetric bending (~1400 cm^{-1}) and ν_1 symmetric stretching (2312 cm^{-1}, 2968 cm^{-1}), as well as the peaks' characteristic for the vibrational mode of water molecules (3640 cm^{-1}) [43]. The IR spectra of the bovine bone precursor included a high-intensity peak at 1008 cm^{-1}, corresponding to ν_3 symmetric stretching of $(PO_4)^{3-}$ groups along with peaks of lower intensity, corresponding to CO_3^{2} groups. The bone organic component is represented by peaks corresponding to amide in collagen at: 1645 cm^{-1} (amide I vibrations), ~1550 cm^{-1} (amide II vibrations) and ~1200 cm^{-1} (amide I vibrations).

After synthesis and thermal treatment, the IR spectra of all three precursors included similar peaks (**Figure 7**). The first peaks, corresponding to ν_4 symmetric bending of $(PO_4)^{3-}$ (563 cm^{-1}, 600 cm^{-1}) are more well defined for the bovine bone-derived material (bovine bone-derived ceramic) in comparison with marble and seashell-derived ones (marble-derived ceramic and

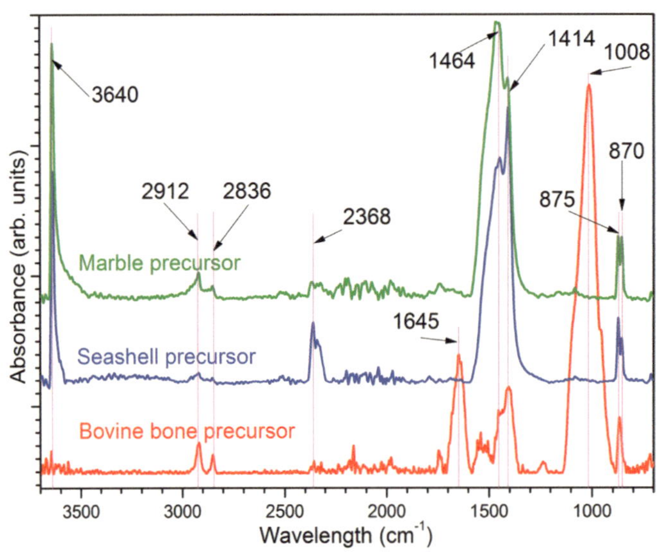

Figure 6. IR spectra of marble, seashell, and bovine bone precursors used for biocompatible calcium phosphate preparation.

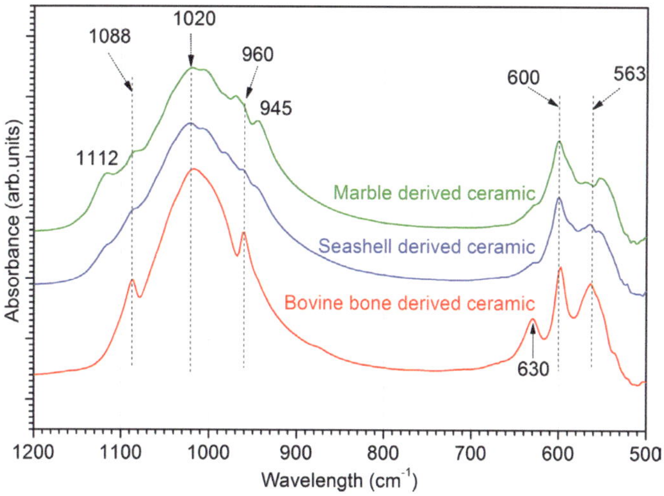

Figure 7. IR spectra of final bioceramic products obtained after indirect chemical synthesis and thermal treatment of marble and seashell precursors (marble-derived ceramic and seashell-derived ceramic, respectively) and thermal treatment at 1000°C of cortical bovine bone (bovine bone-derived ceramic).

seashell-derived ceramic, respectively). Bovine bone-derived ceramic samples also exhibit a peak corresponding to vibration of structural $(OH)^{-}$ groups (630 cm^{-1}), suggesting a higher water content in this samples. Peaks' characteristic to phosphate groups were identified in the 900–1150 cm^{-1} region for all the three types of bioceramics. The IR spectra of bovine bone-derived ceramic includes a well-defined peak at 1020 cm^{-1} with two shoulders at 960 and 1088 cm^{-1}, which resembles well to the characteristic spectra of HA [44]. These peaks are assigned to ν_1 symmetric stretching of $(PO_4)^{3-}$ (960 cm^{-1}) and ν_3 asymmetric stretching of $(PO_4)^{3-}$ (1020, 1088 cm^{-1}). In good agreement with the XRD results (**Figure 5**), the marble-derived bioceramic spectra includes two additional peaks at 945 and 1112 $cm^{-1,}$ corresponding to β-TCP [45], while the seashell-derived bioceramic (seashell-derived ceramic) exhibits a single peak of lower intensity at 1020cm^{-1}, suggesting a lower crystallinity degree of the material with no significant differentiation between the HA and β-TCP peaks.

5. Conclusions and future perspectives

This study proved once more that in the quest of finding an excellent bone substituent, calcium phosphates raised a new level of knowledge due to the generous marine and land bioresources that can be converted. In this context, several drawbacks of current alloplastic methods can be forecasted and minimally invasive surgery shall be needed.

An insightful investigation was carried out in terms of three possible natural precursors—marble, seashell and bovine bone—for biogenic HA synthesis. For the first two, an improved and fully parameterized chemical method was proposed; marble itself serving as an innovative alternative. This led to significant morphological, compositional and structural variations between final stage products. The precursors' structural examination revealed, as it was expected, three polymorphic calcium carbonate forms (calcite, dolomite and aragonite) and a typical bone-like phase composition. Thus, the intermediary thermal treatment affects the initial compact microstructure either by cracking and aeration (marble and seashell) or by transforming the haversian canals and lacunae into size distinctive pores (bovine bone). Further, post-synthesis heat treatment processing constituted a key objective for marble- and seashell-derived powders, which allowed for biphasic powdery calcium phosphates development. Contrary, in case of bone-derived products, there were no structural or compositional events, the final product consisting of pure crystalline HA.

In terms of naturally derived calcium phosphates, future perspectives are mainly correlated to [1] product manufacturing through both the conventional and additive (SFF) methods, [2] controlled porosity for an optimal vascularization and osseointegration and [3] complete standardization for industrial fabrication. In this respect, further thorough research is required.

Acknowledgements

This work was supported by a grant of the Romanian National Authority for Scientific Research and Innovation, CNCS – UEFISCDI, project number PN-III-P2-2.1-PED-2016-0892.

Author details

Florin Miculescu[1]*, Aura-Cătălina Mocanu[1,2], Andreea Maidaniuc[1,3], Cătălina-Andreea Dascălu[1], Marian Miculescu[1], Ştefan Ioan Voicu[4] and Robert-Cătălin Ciocoiu[1]

*Address all correspondence to: m_miculescu@yahoo.com

1 Department of Metallic Materials Science, Physical Metallurgy, Faculty of Materials Science and Engineering, University Politehnica of Bucharest, Romania

2 Research, Development and Innovation Department, S.C. Nuclear NDT Research and Services S.R.L., Bucharest, Romania

3 Destructive and Nondestructive Testing Laboratory, S.C. Nuclear NDT Research and Services S.R.L., Bucharest, Romania

4 Department of Analytical Chemistry and Environmental Engineering, Faculty of Applied Chemistry and Materials Science, University Politehnica of Bucharest, Romania

References

[1] Anastas PT, Warner JC. Principles of green chemistry. Green chemistry. Theory and Practice. 1998:29-56

[2] Mucalo M. Animal-bone derived hydroxyapatite in biomedical applications. In: Hydroxyapatite (HAp) for Biomedical Applications. Elsevier Science. 2015:307-342

[3] Barrère F, van Blitterswijk CA, de Groot K. Bone regeneration: Molecular and cellular interactions with calcium phosphate ceramics. International Journal of Nanomedicine 2006;**1**(3):317

[4] Figueiredo M, Fernando A, Martins G, Freitas J, Judas F, Figueiredo H. Effect of the calcination temperature on the composition and microstructure of hydroxyapatite derived from human and animal bone. Ceramics International. 2010;**36**(8):2383-2393

[5] Herliansyah MK, Muzafar C, Tontowi AE. Natural bioceramics bone graft: A comparative study of calcite hydroxyapatite, gypsum hydroxyapatite, bovine hydroxyapatite and cuttlefish shell hydroxyapatite. Proceedings of the Asia Pacific Industrial Engineering & Management Systems, Bangkok. 2012:1135-1145

[6] Akram M, Ahmed R, Shakir I, Ibrahim WAW, Hussain R. Extracting hydroxyapatite and its precursors from natural resources. Journal of Materials Science. 2014;**49**(4):1461-1475

[7] Vallet-Regi M. Bio-ceramics with Clinical Applications. John Wiley & Sons. Chichester, United Kingdom. 2014:25-57

[8] Dorozhkin SV, Tiwari A, Gerhardt RA, Szutkowska M. Biphasic, triphasic, and multiphasic calcium orthophosphates. Advanced Ceramic Materials. 2016:33-95

[9] Cree D, Rutter A. Sustainable bio-inspired limestone eggshell powder for potential industrialized applications. ACS Sustainable Chemistry & Engineering. 2015;**3**(5):941-949

[10] Komur B, Lohse T, Can HM, Khalilova G, Geçimli ZN, Aydoğdu MO, et al. Fabrication of naturel pumice/hydroxyapatite composite for biomedical engineering. Biomedical Engineering Online. 2016;**15**(1):81

[11] Pon-On W, Suntornsaratoon P, Charoenphandhu N, Thongbunchoo J, Krishnamra N, Tang IM. Hydroxyapatite from fish scale for potential use as bone scaffold or regenerative material. Materials Science and Engineering: C. 2016;**62**:183-189

[12] Mondal S, Pal U, Dey A. Natural origin hydroxyapatite scaffold as potential bone tissue engineering substitute. Ceramics International. 2016;**42**(16):18338-18346

[13] Vecchio KS, Zhang X, Massie JB, Wang M, Kim CW. Conversion of bulk seashells to biocompatible hydroxyapatite for bone implants. Acta Biomaterialia. 2007;**3**(6):910-918

[14] Zhang X, Vecchio KS. Conversion of natural marine skeletons as scaffolds for bone tissue engineering. Frontiers of Materials Science. 2013;**7**(2):103-117

[15] S-C W, Hsu H-C, Y-N W, Ho W-F. Hydroxyapatite synthesized from oyster shell powders by ball milling and heat treatment. Materials Characterization. 2011;**62**(12):1180-1187

[16] Shavandi A, Bekhit AE-DA, Ali A, Sun Z. Synthesis of nano-hydroxyapatite (nHA) from waste mussel shells using a rapid microwave method. Materials Chemistry and Physics. 2015;**149**:607-616

[17] Rujitanapanich S, Kumpapan P, Wanjanoi P. Synthesis of hydroxyapatite from oyster shell via precipitation. Energy Procedia. 2014;**56**:112-117

[18] Brzezińska-Miecznik J, Haberko K, Sitarz M, Bućko MM, Macherzyńska B. Hydroxyapatite from animal bones—Extraction and properties. Ceramics International. 2015;**41**(3, Part B):4841-4846

[19] Niakan A, Ramesh S, Ganesan P, Tan C, Purbolaksono J, Chandran H, et al. Sintering behaviour of natural porous hydroxyapatite derived from bovine bone. Ceramics International. 2015;**41**(2):3024-3029

[20] Macha IJ, Ozyegin L, Oktar FN, Ben-Nissan B. Conversion of ostrich eggshells (*Struthio camelus*) to calcium phosphates. Journal of the Australian Ceramic Society. 2015;**51**(1):125-133

[21] Sunil BR, Jagannatham M. Producing hydroxyapatite from fish bones by heat treatment. Materials Letters. 2016;**185**:411-414

[22] Roy DM, Linnehan SK. Hydroxyapatite formed from coral skeletal carbonate by hydrothermal exchange. Nature. 1974;**247**(5438):220-222

[23] Kumar GS, Girija EK, Venkatesh M, Karunakaran G, Kolesnikov E, Kuznetsov D. One step method to synthesize flower-like hydroxyapatite architecture using mussel shell bio-waste as a calcium source. Ceramics International. 2017;**43**(3):3457-3461

[24] Yang Y, Yao Q, Pu X, Hou Z, Zhang Q. Biphasic calcium phosphate macroporous scaffolds derived from oyster shells for bone tissue engineering. Chemical Engineering Journal. 2011;**173**(3):837-845

[25] Rathje W. Zur Kenntnis der phosphate I: Über Hydroxylapatit. Bodenkunde und Pflanzenernährung. 1939;**12**(1-2):121-128

[26] Azis Y, Jamarun N, Arief S, Nur H. Facile synthesis of hydroxyapatite particles from cockle shells (*Anadara granosa*) by hydrothermal method. Oriental Journal of Chemistry. 2015;**31**(2):1099-1105

[27] Miculescu F, Maidaniuc A, Stan G, Miculescu M, Voicu S, Cîmpean A, et al. Tuning hydroxyapatite particles' characteristics for solid freeform fabrication of bone scaffolds. Advanced Composite Materials. 2016:321

[28] Miculescu F, Maidaniuc A, Stan GE, Miculescu M, Voicu SI, Ciocan LT. Thermal degradation and morphological characteristics of bone products. In: Tiwari A, Raj B, editors. Reactions and Mechanisms in Thermal Analysis of Advanced Materials. Wiley - Scrivener Publishing LLC; Massachusetts, USA. 2015

[29] Mihailescu N, Stan G, Duta L, Chifiriuc MC, Bleotu C, Sopronyi M, et al. Structural, compositional, mechanical characterization and biological assessment of bovine-derived hydroxyapatite coatings reinforced with MgF_2 or MgO for implants functionalization. Materials Science and Engineering: C. 2016;**59**:863-874

[30] Checa AG, Cartwright JH, Sánchez-Almazo I, Andrade JP, Ruiz-Raya F. The cuttlefish *Sepia officinalis* (Sepiidae, Cephalopoda) constructs cuttlebone from a liquid-crystal precursor. Scientific Reports. 2015;**5**:11513

[31] Yan N, Chen X. Don't waste seafood waste: Turning cast-off shells into nitrogen-rich chemicals would benefit economies and the environment. Nature. 2015;**524**(7564):155-158

[32] Kel D, Gökçe H, Bilgiç D, Ağaoğulları D, Duman I, Öveçoğlu M, et al., editors. Production of Natural Bioceramic from Land Snails. Key Engineering Materials. Trans Tech Publ; Switzerland, 2012

[33] Singh A, Purohit K. Chemical synthesis, characterization and bioactivity evaluation of hydroxyapatite prepared from garden snail (helix aspersa). Journal of Biotechnology & Biomaterials. 2011;**1**:104

[34] Wiedemann H-G, Bayer G. Note on the thermal decomposition of dolomite. Thermochimica Acta. 1987;**121**:479-485

[35] Prabakaran K, Balamurugan A, Rajeswari S. Development of calcium phosphate based apatite from hen's eggshell. Bulletin of Materials Science. 2005;**28**(2):115-119

[36] Miculescu F, Stan GE, Ciocan LT, Miculescu M, Berbecaru A, Antoniac I. Cortical bone as resource for producing biomimetic materials for clinical use. Digest Journal of Nanomaterials and Biostructures. 2012;**7**(4):1667-1677

[37] Liu Q, Huang S, Matinlinna JP, Chen Z, Pan H. Insight into biological apatite: Physiochemical properties and preparation approaches. BioMed Research International. 2013; **2013**:1-13

[38] Pramanik S, Pingguan-Murphy B, Cho J, Osman NAA. Design and development of potential tissue engineering scaffolds from structurally different longitudinal parts of a bovine-femur. Scientific Reports. 2014;**4**:1-10

[39] Beckett S, Rogers KD, Clement JG. Inter-species variation in bone mineral behavior upon heating. Journal of Forensic Science. 2011;**56**(3):571-579

[40] Miculescu F, Mocanu A-C, Dascălu CA, Maidaniuc A, Batalu D, Berbecaru A, et al. Facile synthesis and characterization of hydroxyapatite particles for high value nanocomposites and biomaterials. Vacuum. 2017 (in press). https://doi.org/10.1016/j.vacuum.2017.06.008

[41] Miculescu F, Mocanu AC, Stan GE, Miculescu M, Maidaniuc A, Cîmpean A, et al. Influence of the modulated two-step synthesis of biogenic hydroxyapatite on biomimetic products' surface. Applied Surface Science. 2017 (in press). https://doi.org/10.1016/j.apsusc.2017.07.144

[42] Rodriguez-Navarro C, Ruiz-Agudo E, Luque A, Rodriguez-Navarro AB, Ortega-Huertas M. Thermal decomposition of calcite: Mechanisms of formation and textural evolution of CaO nanocrystals. American Mineralogist. 2009;**94**(4):578-593

[43] Khiri MZA, Matori KA, Zainuddin N, Abdullah CAC, Alassan ZN, Baharuddin NF, et al. The usability of ark clam shell (*Anadara granosa*) as calcium precursor to produce hydroxyapatite nanoparticle via wet chemical precipitate method in various sintering temperature. SpringerPlus. 2016;**5**(1):1206

[44] Markovic M, Fowler BO, Tung MS. Preparation and comprehensive characterization of a calcium hydroxyapatite reference material. Journal of Research of the National Institute of Standards and Technology. 2004;**109**(6):553

[45] Jillavenkatesa A, Condrate Sr R. The infrared and Raman spectra of β-and α-tricalcium phosphate ($Ca_3(PO_4)_2$). Spectroscopy Letters. 1998;**31**(8):1619-1634

Chapter 7

Hydroxyapatite-Based Materials for Potential Use in Bone Tissue Infections

Katarzyna Szurkowska, Aleksandra Laskus and Joanna Kolmas

Additional information is available at the end of the chapter

http://dx.doi.org/10.5772/intechopen.71604

Abstract

Hydroxyapatite materials, due to their high biocompatibility, play a crucial role in orthopaedics and bone surgery as alternatives to autologous bone grafts. It was also found that coatings of metallic implants with hydroxyapatite layer improve significantly their osseointegration. Due to its bioactivity, osteoconductivity and non-toxicity, hydroxyapatite is also widely used as a component of hybrid biomaterials. The implantation of "foreign" materials brings one major concern that is the risk of potential bone tissue infections or chronic osteomyelitis. In turn, the main problem concerning bacterial infection treatment is to obtain an adequate, bactericidal drug concentration maintained for a sufficient period of time in the bone tissue. Therefore, recent developments of materials engineering are focused on delivery antibiotics directly into the affected bone. To achieve this goal, hydroxyapatite-based materials are frequently studied as carriers for antibacterial drugs. For effective support of antibiotic therapy, the antibacterial activity of certain ions (including silver, zinc or copper) may be applied. In our work, we present recent developments on ceramic materials for bacterial bone infections: hydroxyapatite-based carriers for antibiotics and modifications of hydroxyapatite with antibacterial ions. In this review, state-of-the-art and current applications of such materials are presented and discussed. We want to also present our recent results.

Keywords: hydroxyapatite, drug delivery, antibiotics, ionic substitution, antibacterial properties

1. Introduction

Hydroxyapatite (HA) is a material widely used in regenerative medicine, bone and dental surgery, conservative dentistry as well as implantology [1, 2]. HA resembles the main inorganic component of mineralized tissues (biological apatite), which in combination with its non-toxic and, most importantly, osseoconductive properties makes it an asset for biomaterial engineering [3]. HA is considered to be the gold standard in bone tissue regeneration. In clinical practice, it is used in the form of powders or granules as filler for bone replacement or for repair of post-resection defects [4, 5]. HA is also successfully used as a coating material for metallic implants due to its bioactivity and favourable effects on the osseointegration process [6]. Porous structures may be used as temporary scaffolds for newly formed osseous tissue. In dentistry, HA is a component of dental materials such as dental cements and toothpastes [7]. Moreover, it has further uses in polymer/ceramic bone composite materials, not only as a bioactive material but also as a provider of desirable mechanical properties [8, 9]. Current research on HA bioceramics is conducted with a view to achieve two main goals: (1) to improve the biocompatibility of synthetic HA and (2) to provide synthetic HA with additional biological properties. The first goal can be achieved using partial ionic modification of synthetic HA. It should be stressed at this point that biological apatite is not pure hydroxyapatite, it is carbonated hydroxyapatite with a considerably reduced content of calcium and structural hydroxyl groups [10]. It also contains a number of various ions, primarily magnesium (Mg^{2+}), but also sodium (Na^{+}), potassium (K^{+}), zinc (Zn^{2+}), manganese (Mn^{2+}), silicate (SiO_4^{4-}) and hydrogen phosphate (HPO_4^{2-}). The "foreign ions" incorporated into the structure of HA contribute significantly to its properties such as the size of single crystals, agglomeration tendency and solubility.

New biological properties of HA may also lead to its enrichment with additional ions. For example, the introduction of strontium ions (Sr^{2+}) provides HA with antiresorptive properties, as the strontium ions have an inhibiting effect on the activity of osteoclasts, while also stimulating osteoblasts [11]. HA material containing selenites (SeO_3^{2-}) may be used in turn in bone tumour therapy [12]. Commercially available apatite material enriched with silicon ions (Actifuse®) contributes positively to osteogenesis by promoting the formation of bone and its natural remodelling [13].

Upgrading HA materials may be achieved using physical or chemical binding of drugs. Therefore, recent research on HA bioceramics focused on producing multifunctional materials, which, in addition to being used as scaffolds for growing tissue, could also release drugs directly into the bone in the affected area [14]. The literature describes research on HA as a delivery system for antiresorptive (e.g., bisphosphonates) and anticancer drugs (e.g., doxorubicin and cisplatin), as well as antibiotics mainly against perioperative and intraoperative infections [15–17].

This chapter presents so far achievements in the field of HA materials for bone tissue infections (see **Figure 1**). In addition to antibiotic delivery systems, herein the focus will be put on HA modified by ions with proved antibacterial activity. Further on, opportunities for developing multifunctional HA-based materials for applications related to prevention and treatment of bone infections will be discussed.

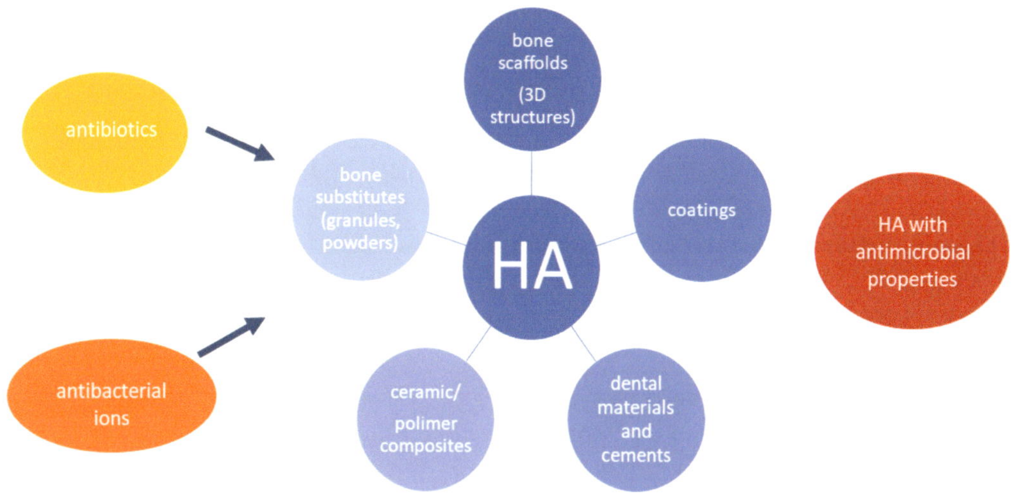

Figure 1. Scheme illustrating the main applications of HA and the ways to obtain antibacterial properties.

2. Hydroxyapatite-based antibiotic delivery systems

Bone tissue infections are one of the most frequently occurring side effects of bone surgeries. Such a complication may lead to severe bone loss, implant failure or even amputation [14]. Osteomyelitis, periodontitis and spondylodiscitis are important bone tissue infections [18, 19]. They are most commonly caused by infectious isolates of G-positive bacteria, such as *Staphylococcus aureus* and *Streptococcus* spp.; G-negative bacteria: *Salmonella* spp., *Mycobacterium tuberculosis, Pseudomonas aeruginosa*; and fungi: *Candida* spp. The treatment of bone infections meets several serious clinical problems. Usually, antibiotic therapy involves 3 weeks of oral treatment followed by 3 weeks of intravenous therapy [19]. Bone tissue is poorly vascularized; thus, the antibiotic doses must be high enough to reach prolonged antibacterial concentration at the infected site. This high dosage of antibiotics may cause systemic toxic effects like nephrotoxicity, ototoxicity, hepatotoxicity, allergy or gastrointestinal syndromes [14].

Despite long, high-dose therapies, standard treatments of bone infections are still not effective enough. Due to the problems mentioned above, drug delivery systems targeting bones have been developed. The material frequently chosen as the system matrix is hydroxyapatite (HA).

Due to its porosity, HA may provide proper loading and long-term release of antibacterial agents, which is crucial for the antibacterial effectiveness of such a system. However, its poor mechanical properties (brittleness) have led scientists to combine pure HA with natural or synthetic polymers. Gelatine [20, 21], alginates [22–25], chitosan [25–27], collagen [28–30], polyvinyl alcohol (PVA) [31–33], polyacids [34–40] and cyclodextrins [41–43] are frequently used to improve not only the properties mentioned above but also the stickiness of fabricated composite scaffolds, microspheres, etc. Thus, investigations into drug delivery systems loaded with antibiotics include the use of HA alone [44–63] and HA accompanied by other substances [20–40, 42, 43, 64–66].

The most frequently used antibiotics in local drug delivery systems are vancomycin (VAN) [18, 20, 21, 27, 33, 42, 48, 53, 60–63, 65] and gentamicin (GT) [18, 23, 25, 27, 35, 46, 57–59, 65]. These are also the most ubiquitously applied antibacterial agents in systemic therapy of bone tissue infections. Herein, the examples of antibiotic delivery systems based on HA and loaded with VAN or GT will be presented.

2.1. Vancomycin

Vancomycin (VAN) is used to treat methicillin-resistant *Staphylococcus aureus* (MRSA) infections in bone. The drug is administered parenterally; however, poor vascularization of bone tissues may cause insufficient local concentration of the antibiotic. Furthermore, severe side effects, such as ototoxicity and nephrotoxicity, are driving investigations into local delivery systems for VAN.

In one study [62], different materials characterized by various pHs were used to incorporate VAN. Namely, the investigations were focused on brushite cement (pH = 2.4), HA cement (pH = 9.4) and apatite xerogel (pH = 7.4). The influence of pH on the antibiotic release mode was analysed. The outcomes of the experiment revealed that pH affected the release kinetics. Despite the fact that the eluent from apatite cement exceeded the minimum inhibitory concentration (MIC), the system based on this material was ineffective against *S. aureus*. Yang et al. [27] covered metallic implants of bone with a chitosan/vancomycin composite. The composite's components were interconnected with hydrogen bonding. The electrochemical deposition technique was employed to cover the implant with a layer of composite. Next, the additional, external HA layer was placed on the implant. The kinetics of the antibiotic release from both type coatings were then compared. The kinetics showed that chitosan coating resulted in an impressive initial burst of a drug compared with the chitosan/HA composite. It may be concluded that the addition of HA has a significant impact on the prolonged release of the antibiotic.

The antibacterial activity of HA-based VAN-loaded delivery systems is usually examined *in vitro*. However, some studies involve *in vivo* tests to investigate the antibacterial effectiveness of fabricated systems. Joosten et al. [61] tested the antibacterial activity of VAN-loaded HA cement in *S. aureus*–induced chronic osteomyelitis. The infection was induced in the tibia of New Zealand white rabbits. The HA cement was an effective VAN carrier even for the treatment of MRSA.

Lian et al. [31] tested HA/collagen/calcium sulphate composites loaded with VAN also in rabbits. Bone infection was induced in the condyle lateralis femoris. After 12 weeks of implantation, micro-CT graphs have shown an excellent bone reconstruction with implants containing VAN (see **Figure 2**).

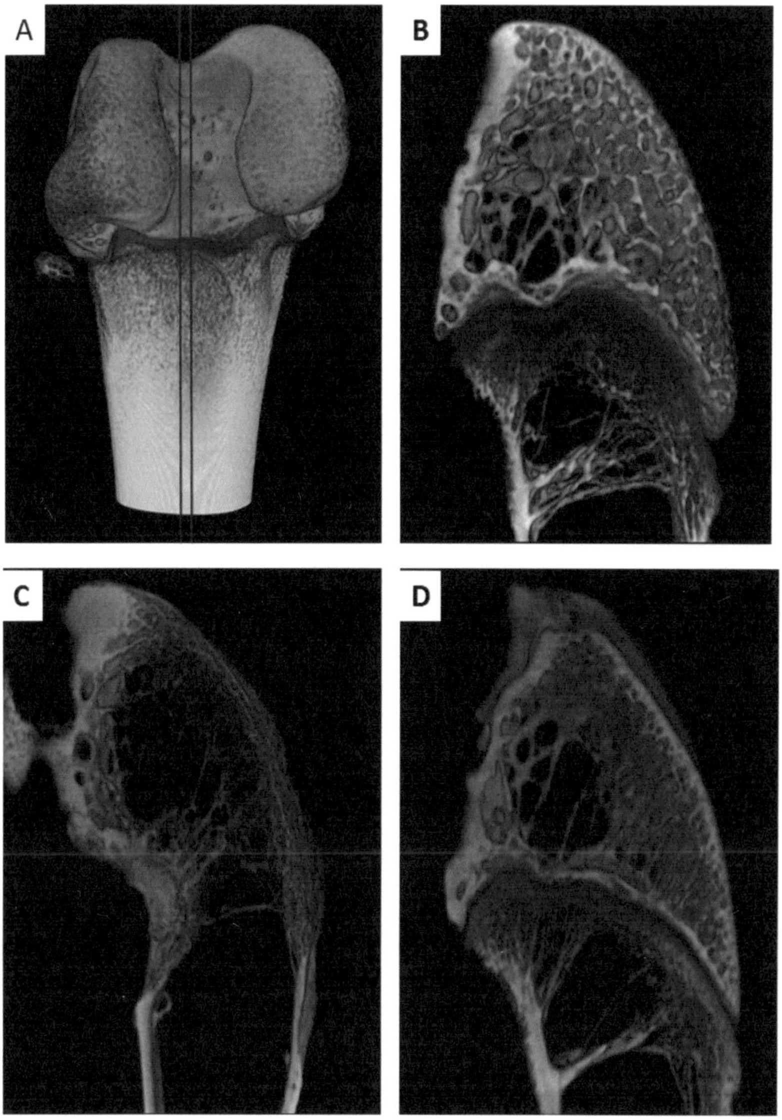

Figure 2. Micro-CT graphs taken 12 weeks after focal debridement. (a) Cross-section position (red line), (b) normal bone, (c) nHAC/CSH group, and (d) VCM/nHAC/CSH group. Abbreviations: nHAC/CSH – nanohydroxyapatite/collagen/calcium sulphate composite VCM/nHAC/CSH – nanohydroxyapatite/collagen/calcium sulphate composite loaded with vancomycin. Reprinted from Ref. [31], the open access article distributed under the Creative Commons Attribution License.

Some commercial materials were also tested for their effectiveness as the matrices of antibiotic drug delivery systems [18, 53, 65]. Interesting outcomes were found by Rauschmann et al. [65] who compared PerOssal® and calcium sulphate (CS) as drug loading matrices. PerOssal® is a biodegradable composite consisting of nano-sized HA and CS. The pellets synthesized from both materials were soaked in two antibiotics: VAN and gentamicin. The release of the drugs from the materials was studied. Surprisingly, PerOssal® demonstrated a higher initial release and a lower release of VAN after approximately 5 days, while in the case of gentamicin, the release mode from the materials exhibited no significant difference.

2.2. Gentamicin

Gentamicin (GT) is a broad-spectrum antibiotic from the group of aminoglycosides. It is mainly used in infections involving Gram-negative bacteria (i.e. *Pseudomonas* and *Enterobacter* spp.). Due to poor oral absorption, GT is commonly administered by injection. GT is frequently used as a model, antibacterial agent in HA-based drug delivery systems. Guo et al. [57] examined the influence of the HA's porosity on GT's loading. Mesoporous, carbonated HA microspheres exhibited a higher drug loading efficiency of 70–75% more than the conventional HA particles. It is important to note that the hierarchical nanostructure with developed meso- and microporosity allowed for an efficient loading of drug and, at the same time, a slow and sustained release of GT.

The association between porosity and drug loading was also studied by other researchers. To synthesize porous HA microspheres, the ice-template spray drying (ITSD) technique was applied by Yu et al. [34]. Drug loading efficacy increased with the increase of the porosity of the HA microspheres. Additionally, the transformation of the structure of the pores from cellular and independent ones to three-dimensional interconnected pore networks had a significant impact on the initial burst of the drug.

A hybrid material containing HA and covalently coated, hardly degradable keratin was described in Ref. [66] as an innovative system for GT delivery. The presence of keratin resulted in a greater immobilization of the antibiotic compared with HA/gelatine material. Moreover, the hybrid was non-toxic and stimulated osteoblast proliferation. It is important to note that the sustainable, prolonged GT release provided efficient antibacterial activity for at least 120 days.

2.3. Other antibiotics used in HA-based delivery systems

Among the other antibiotics applied as the model drugs in the systems targeting bones, penicillins, mainly amoxicillin [22, 37, 38, 44–46], cephalosporins [44, 46, 47], fluoroquinolones [33, 42, 43, 48, 49], including ciprofloxacin [33, 42, 43, 48, 49] and tetracyclines [28, 36, 51, 52], should be mentioned. In some studies, aminoglycosides (tobramycin and amikacin) [44, 46, 54], erythromycin (macrolides) [22, 32], tigecycline (glycylcyclines) [55, 56], linezolid (oxazolidinones) [64], rifapentine (ansamycin-like antibiotic) [41], clindamycin (lincosamides) [39, 40, 50], chloramphenicol [45] or chlorhexidine [67] (a bactericidal and bacteriostatic agent, not classified as an antibiotic) were used. The most interesting investigations concern loading more than one antibiotic into the same material [22, 42, 44–46, 48, 49].

Stigter et al. [44] compared the efficacy of the incorporation of different antibiotics into carbonated HA coatings on titanium implants. The outcomes showed that the incorporation rate depends on the chemical structure of the drug. Antibiotics that contained a carboxylic group, such as cefalotin, carbenicillin or cefamandole, were better incorporated than the others. In addition, these drugs exhibited a slower release from HA coatings.

In turn, Ferraz et al. [22] loaded nanohydroxyapatite microspheres with amoxicillin, amoxicillin + clavulanic acid and erythromycin. Two types of microspheres, with varied porosity, were tested. The release profile from both types of microspheres consisted of a fast initial release followed by long-term sustained release. The microspheres with higher porosity and a greater surface area released more antibiotic during the first days. The antibacterial activity was tested against *S. aureus* and *Escherichia coli*. The obtained results have shown that the materials exhibited good, long-term antimicrobial activity.

Detailed study focused on HAs with controlled porosity and loaded with three antimicrobial agents (vancomycin, ciprofloxacin and gentamicin) were described in Ref. [48]. It was concluded that the adsorption of antibiotics was significantly higher in microporous HA than in crude dense discs. Moreover, the amount of adsorbed VAN was significantly higher than ciprofloxacin and gentamicin. Exposure to different bacteria species such as *S. aureus, Staphylococcus epidermidis and E. coli* demonstrated efficient antibacterial activity for all the materials. However, the microporosity of HA disc significantly prolonged the release of antibacterial agents.

A very interesting research was presented by Ghosh et al. [49]. HA cements were prepared with two types of nanohydroxyapatites and loaded with ciprofloxacin or VAN. Self-setting time reactions were controlled using the different weight ratios of the nanohydroxyapatites and had an impact on the release rate of antibiotics. The results have shown that, with modification of cement components, tuneable antibiotic release rates may be obtained. The biological tests presented good biocompatibility and non-toxicity to osteoblastic and osteoclastic cells.

The possibility of efficient fast loading of antibiotics in HA was studied by Brohede et al. [46]. The HA coatings on titanium implants were loaded with tobramycin, gentamicin, amoxicillin or cefalotin via soaking for varying periods of time (15 mins to 24 h). The results of antibacterial tests have shown that even the shortest loading time was sufficient to release enough drug for the next 24 h and inhibit bacterial growth.

3. Hydroxyapatite materials doped with antibacterial ions

The antibiotic resistance demonstrated by many bacterial species has stimulated attempts to produce new materials with efficient antibacterial properties. It is also important to note that implant-related/bone infections are caused by bacterial adhesion and biofilm formation. Biofilms are difficult to treat with standard antibiotic therapy. Thus, searching for new antibacterial strategies seems to be justifiable. As was mentioned above, HA doped with functional ions (i.e. Ag^{+}, Zn^{2+}, Cu^{2+}, SeO_3^{2-}) may be applied for perioperative and intraoperative prevention and treatment of bone infections.

3.1. Silver-substituted hydroxyapatite

Silver exhibits a wide spectrum of actions against bacteria, viruses and fungi with a relatively low risk of resistance developing [68]. Silver compounds are effective against some common pathogens such as *E. coli, S. aureus* and *S. epidermidis* and, more importantly, methicillin- and vancomycin-resistant *S. aureus* (MRSA and VRSA) [69–72]. Other susceptible microorganisms include *Klebsiella pneumoniae, Providencia stuartii, Citrobacter freundii, Micrococcus luteus, P. aeruginosa, Pneumococcus* spp., *Streptococcus mutans, Aggregatibacter actinomycetemcomitans, Porphyromonas gingivalis* as well as yeasts *Issatchenkia orientalis* and *C. albicans* [73–76].

The main mechanism of action consists of the inactivation of microbial proteins through interactions with thiol groups (–SH) and the formation of inactive S-Ag bonding. Silver also affects bacterial DNA, precluding its replication. Another mechanism includes increased reactive oxygen species (ROS) production, leading to abnormally high permeability of microbial cells [68].

Silver-substituted HA (Ag-HA) can be obtained using several main synthesis methods, such as wet precipitation (using salts [74] or the neutralization reaction [77]), sol-gel technique [71, 78], hydrothermal method [79], electrochemical deposition [80] and magnetron sputtering [73]. Additional treatment includes sintering [81] or microwave assistance [72]. A wide range of silver substitutions have been investigated-from ultra-trace amounts such as 0.04 ppm [79] or 0.002 mole Ag per 1 mole HA [77] up to 10 wt.% [82, 83]. To better evaluate the relationship between silver concentration and physicochemical properties and the biological activity of Ag-HA samples, studies usually include a series of samples with various Ag contents.

The antibacterial activity of silver is dose-dependent and increases with higher silver concentrations. However, higher doses of silver increase the risk of severe cytotoxic effects to mammalian cells. HA with 10 wt% of silver was synthesized by Nath et al. [82] via the sintering of mechanically mixed powders at 1200°C. Biocompatibility was confirmed on mouse fibroblast (L929) and human osteosarcoma (MG-63) cells. Rajendran et al. [83] also confirmed >80% viability of NIH3T3 cells cultured on HA with 10 wt% Ag, but even 3 wt% Ag was sufficiently effective against *S. aureus*. However, Ag-HA nanocomposite coatings on Ti implants with 5 wt% content of metallic Ag exhibited cytotoxic effect on mice osteoblasts, while 2 wt% of Ag was both cytocompatible and inhibited growth of *S. aureus* [84]. These results are consistent with research by Yan et al. [80], where Ag^+-substituted HA coatings with 2.03 wt% of silver exhibited optimal osteogenic and antimicrobial properties. According to Shi et al. [79], the optimal doping concentration of Ag ranges from 0.27 to 2.2 ppm. Lu et al. [85] also emphasized the importance of incorporating an adequate amount of the element to balance antibacterial activity and biocompatibility. Interestingly, heat treatment enhanced biocompatibility without decreasing antimicrobial properties. Another study indicated improved antibacterial activity against *S. aureus, K. pneumoniae* and *C. albicans* after thermal treatment at 600 and 1000°C [81].

Lee et al. [86] prepared nanocomposite fibres composed of Ag-doped HA and polyamide 6. Ag^+ ions were loaded through the ion-exchange mechanism. HA was synthesized in agarose and ethanol medium to obtain the desired properties. Such composites exhibited excellent antimicrobial activity against *K. pneumoniae* and *E. coli* while being slightly less effective against *S. aureus*. Further modification of the antibacterial fibre could extend the application field of Ag-HA, so far predominantly used in hard tissue injuries, to the treatment of skin diseases.

In Ref. [75], HA powders enriched in silver ions were used as coatings on a silicon previously covered with an elastomer, polydimethylsiloxane (PDMS). The antimicrobial activity was measured against *E. coli*, *S. aureus* and *C. albicans* strains. The obtained layers successfully inhibited microbial growth after 24 h of test (see **Figure 3**). Other polymer-based composites with polyvinyl alcohol [71], polyethylene glycol [78] and chitosan [87] were also examined.

Novel nanoscaffold biomaterials, based on porous HA, polyamide 66, titanium dioxide (TiO_2) and various concentrations of Ag^+ ions, were developed and thoroughly examined by Lu et al. [88]. Therapeutic effects of the biomaterial were tested *in vivo* on a large cohort of rabbits with osteomyelitis for 12 weeks. The treatment was successful, scaffolds exhibited both antimicrobial and anti-inflammatory effects and, in addition, stimulation of osteogenesis was observed. *In vivo* silver concentrations following implantation were under toxic levels and no failure of liver or kidney functions occurred.

Titanium discs coated with thermal sprayed Ag-HA (0.5–3.0 wt%) were tested *in vitro*, revealing a reduced ability of biofilm formation by a methicillin-resistant *S. aureus* strain. The effect was confirmed *in vivo* on rats with an MRSA-inoculated 3% Ag-HA disc implanted hypodermic for 7 days. No skin disorder (such as argyria) or wound healing complications were observed [89]. The reduction of viable MRSA by Ag-based coating on tibia implants was also

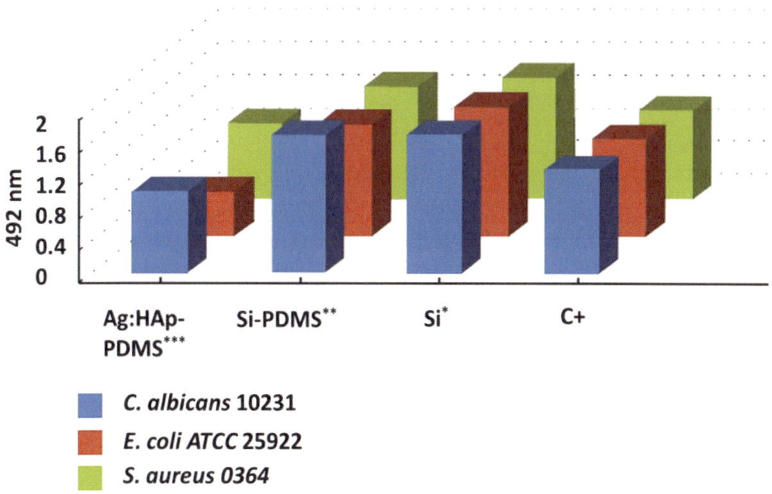

Figure 3. The graphic representation of the microbial activity of *S. aureus* 0364, *E. coli* ATCC 25922 and *C. albicans* 10,231 on Ag:HAp-PDMS layers on Si substrate, PDMS layers on Si substrate (Si-PDMS), and Si substrate (Si) at 48 h. *silicon substrate, **silicon substrate previously coated with PDMS, and ***Ag:HAp nanoparticles on a silicon substrate previously coated with a PDMS layer. Reprinted from Ref. [75], the open access article distributed under the Creative Commons Attribution License.

indicated *in vivo* on rat models [70]. An interesting study by Xie et al. [87] concerned the successful doping of bone morphogenetic protein 2 (BMP-2) into a nanosilver/hydroxyapatite/chitosan composite, which was then implanted into the femurs of rabbits. Favourable bone formation and antibacterial properties were demonstrated *in vivo*.

In 2016, the first clinical study was conducted on 20 human patients with total hip arthroplasty, in which a silver oxide (Ag_2O)-HA implant coating was used. The highest reported silver blood level following the surgery was far below the toxic level. For 1 year after surgery, no significant adverse reactions were observed and the coating prevented postoperative infection [90].

A popular strategy to further improve the properties of Ag-HA is to co-substitute additional ions. The most frequently studied combinations concern the addition of silicate SiO_4^{4-} ions (to improve osteogenic properties) [91] and strontium (Sr^{2+}) ions (to reduce silver cytotoxicity and boost antibacterial properties) [92]. Recently, Aksakal et al. [93] examined multiple HA substitutions with silver, zirconia and yttria, while Kolmas et al. [94] indicated that co-substitution of Ag-HA with carbonate (CO_3^{2-}) ions increased the solubility of samples, thus exhibiting greater antibacterial effect.

3.2. Zinc-substituted hydroxyapatite

Zinc (Zn^{2+}) ion substitution in biomaterials has been thoroughly investigated, for both its osteogenic [95] and antibacterial activities [96–104]. The mechanism of inhibition of microbial growth by zinc ions includes several aspects. Zn^{2+} ions cause damage to cell membranes by bonding with functional groups and increasing the permeability of cells. Moreover, zinc interacts with bacterial enzymes (such as ATPase, glycolytic enzymes or pyruvate kinase), disturbing their correct functionality [98, 103].

Recently, researchers have conducted in-depth investigations into the antibacterial action of zinc-substituted hydroxyapatites (Zn-HA). Samples with various levels of Zn^{2+} substitution were synthesized via the most common methods, namely co-precipitation [100, 102], ion exchange [99], sol-gel [104] and hydrothermal synthesis [101]. Anwar et al. [98] proposed a novel technique: continuous plastic flow synthesis (CPFS), which enables rapid production of HA nanocrystals with a high surface area. Electrospinning of fibres [96] and synthesis mediated by surfactant addition TritonX-100 [103] were also examined.

Common human pathogens used for testing antimicrobial activity were *S. aureus*, *E. coli* and *P. aeruginosa* [98, 100–102]. Individual works concerned the impact of Zn-HA on the growth of *Bacillus subtilis*, *Enterobacter aerogenes*, *Aggregatibacter actinomycetemcomitans*, *Fusobacterium nucleatum*, *S. mutans*, *Shigella flexneri*, *M. luteus*, *Bacillus cereus*, *Porphyromonas gingivalis* and yeast *C. albicans* [97–99, 101–104]. Biocompatibility of Zn-HA biomaterials was demonstrated *in vitro* on human osteoblast-like cells MG-63 [104], human adipose-derived mesenchymal stem cells (MSCs) [102], rat primary osteoprogenitor cells and fibroblast cells MRC-5 [101]. In some studies, the viability rate was better for Zn-containing samples than for pure HA [101, 102].

Thian et al. [102] proved that the addition of Zn^{2+} ions stimulated the bioactivity of HA, since the increased growth of MSC cells, as well as elevated expression of collagen type I and osteocalcin, was observed in case of Zn-HA (1.6 wt% of Zn). Moreover, Tank et al. [103] indicated no significant haemolytic activity of Zn-HA on human blood. Bioactivity *in vitro* was proved by the ability of Zn-HA to form apatite crystals on samples soaked in simulated body fluid (SBF), which increased as the concentration of Zn raised.

Some research provides a comparison of antimicrobial activity against several pathogens. Radovanović et al. [101] investigated the inhibition of growth of *E. coli, S. aureus, P. aeruginosa* and *C. albicans* caused by Zn-HA samples (0.2 and 0.4 mol%) and undoped HA. It was found that sintering the apatites at 1200°C, which led to partial decomposition to more soluble α-TCP, improved antibacterial activity of samples. All tested microorganisms were susceptible to Zn-HA and the degree of reduction increased with higher content of zinc ions.

Slightly different results were reported by Tank et al. [103] who focused on *P. aeruginosa, S. flexneri, M. luteus, S. aureus* and *B. cereus*. Zn substitution ranged from 1.3 wt% to 4.8 wt%. *S. aureus* was the most sensitive strain, even to undoped HA. *M. luteus* was also highly susceptible to Zn-HA samples, while both *B. cereus* and *S. flexneri* exhibited a moderate reduction in the number of colonies. In contrast, Zn-HA samples were ineffective against *P. aeruginosa*.

Several studies indicated that Zn-HA–based materials could also be suitable for the treatment of oral cavity bacterial infections. Zn-HA was effective in inhibiting the growth of common oral pathogenic strains, namely *Aggregatibacter actinomycetemcomitans, Fusobacterium nucleatum* and S. *mutans* [99]. Zn-HA coating on titanium implants demonstrated antibacterial properties against *Porphyromonas gingivalis*, a major cause of chronic periodontitis [104]. An additional advantage of Zn-HA used as an additive to toothpaste is protection against acid enamel erosion [105].

It should be noted that zinc is a popular dopant in multiple substituted HAs. Commonly examined combinations include Zn-HA with Ag^{+} [106] or Cu^{2+} [107] ions used to boost antibacterial activity, and Mg^{2+} [108], SiO_4^{4-} [109], Sr^{2+} [108] or F^{-} ions [110] for additional stimulation of the mineralization process.

3.3. Copper-substituted hydroxyapatite

The antimicrobial activity of copper is linked to its interaction with bacterial proteins, membranes and nucleic acids. An extensive review of antimicrobial applications of copper in the environment was provided by Vincent et al. [111].

Li et al. [112] synthesized copper-substituted HA (molar rate of Cu^{2+}/Ca^{2+} up to 0.15) via ion exchange wet chemical reaction. Obtained materials exhibited a high antibacterial effect against *E. coli*, which increased with the concentration of Cu^{2+} ions. Unfortunately, all Cu-HA samples were cytotoxic to human foetal osteoblast (hFOB) cell lines.

Sahithi et al. [113] combined copper-soaked HA with polyethylene glycol 400 (PEG 400) to further extend its antimicrobial activity. Cu-HA exhibited antibacterial activity against *E. coli* and *S. aureus*, the effect of which increased against *S. aureus* after combination with PEG 400. MTT assay carried out on rat primary osteoprogenitor cells indicated cytocompatibility of the samples.

Antimicrobial activity of Cu-HA as well as Cu-FA (copper-substituted fluorapatite) was tested against *E. coli, S. aureus* and *C. albicans* [114]. The increase of copper substitution in hydroxyapatite enhanced activity against *S. aureus* and *C. albicans*, but Cu-HAs were not active enough against *E. coli*. Cu-FA was effective against all tested microorganisms with increasing activity in the following order: *C. albicans* < *S. aureus* < *E. coli*. Cu-FA may be more effective due to the release of fluoride ions. The same pathogen strains were used in Ref. [115], where the antibacterial activity of Cu-HA was compared with results for Zn-HA.

Radovanović et al. [116] compared Ag^+- and Cu^{2+}-substituted biphasic materials, based on ion-doped HA and α-TCP, obtained after annealing monophasic-substituted HA samples at 1200°C. For antimicrobial tests *in vitro, S. aureus, E. coli, P. aeruginosa* and *C. albicans* were used. Antimicrobial activity increased with the increase of ionic concentration. The activities of biphasic materials were very high and comparable in the case of Ag^+ and Cu^{2+} substitution. The only difference was observed in monophasic Cu-HA against *C. albicans*, which was much less effective, especially with the smaller concentration of Cu^{2+} ions. *In vitro* biocompatibility was demonstrated on MRC-5 human fibroblast cells, but it should be noted that the addition of Cu^{2+} ions slightly reduced the viability of cells.

3.4. Selenium-substituted hydroxyapatite

Tran et al. [117] confirmed antibacterial properties of cellulose discs coated with organoselenium-methacrylate polymer against *P. aeruginosa* and *S. aureus*. 0.2 wt% of selenium completely inhibited bacterial attachment, growth and formation of a biofilm. Strong activity of selenium nanoparticles against *S. aureus* was confirmed by Tran and Webster [118]. These studies led to more research concerning the antimicrobial activity of selenium-based hydroxyapatite (Se-HA) [119–121]. Rodriguez-Valencia et al. [119] fabricated selenium-substituted carbonated HA coatings by the pulsed laser deposition method. Samples contained selenium in the form of selenite ion SeO_3^{2-}. Coatings prevented the formation of biofilms by *P. aeruginosa* and *S. aureus* strains and reduced the number of colony-forming units (CFUs). Uskoković et al. [120] compared Se-HA obtained by co-precipitation and ion-exchange sorption methods. Selenite contents ranged from 0.3 to 3 wt% and the precipitation synthesis was about 10 times more effective in introducing selenium. Se-HA samples were strongly effective against *E. coli* and *S. aureus*, while being less effective against *Salmonella enteritidis* and ineffective against *P. aeruginosa*. Similar results were obtained by Kolmas et al. [121]. **Figure 4** illustrates significant bacterial growth inhibition caused by selenite anions. Moreover, selenium content was in correlation with the reduction of the viability of mouse osteosarcoma cells, and the induction of apoptosis was selective, without reducing the viability of fibroblast cells. Se-HA also exhibited osteoinductive effect by increasing the gene expression of pre-osteoblastic MC3T3-E1 cells. These promising results mean that selenium substitution in hydroxyapatite will probably get more popular in upcoming years.

In addition to the well-known elements with well-established antibacterial activity, some less popular elements for such a combination, like cerium, gallium, cobalt and strontium, should be mentioned [122].

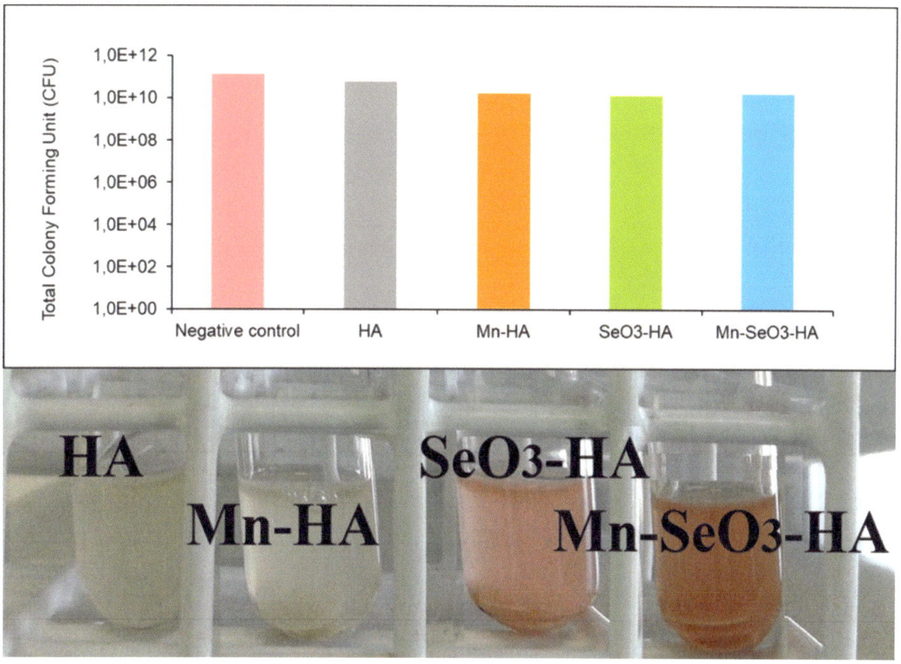

Figure 4. Antibacterials activity of the materials on *S. aureus*. Reprinted from Ref. [121], the open access article distributed under the Creative Commons Attribution License.

4. Hydroxyapatite with antibacterial ions and loaded antibiotics

Some hydroxyapatites enriched with antibacterial ions have been used to create systems containing antibiotics. Most of these systems have been developed using silver ions because of their strong antibacterial properties.

Ivashenko et al. [123] investigated the effect of silver ions in HA structure on the adsorption rates of ciprofloxacin. The research was carried out using commercially available materials such as Biomin G® (HA) and Biomin GIS® (HA enriched in an Ag^+ amount of <0.1 wt%). Interestingly, the presence of silver ions in HA led to lowered specific surface area and significantly decreased adsorption rates of ciprofloxacin when compared with undoped material. Unfortunately, no research was done to test the antibacterial activity or release of silver ions or ciprofloxacin.

Another work [124] proposes long HA nanowires enriched with Ag^+ ions and ciprofloxacin. The material performed high and long-termed effectiveness against *E. coli* and *S. aureus*.

Ciprofloxacin and tetracycline were also adsorbed on a thin film made of Ag-HA [125]. *In vitro* microbiological tests have shown that thin films containing Ag-HA and selected antibiotics may become an effective solution in the prevention and treatment of bone infections (see **Figure 5**).

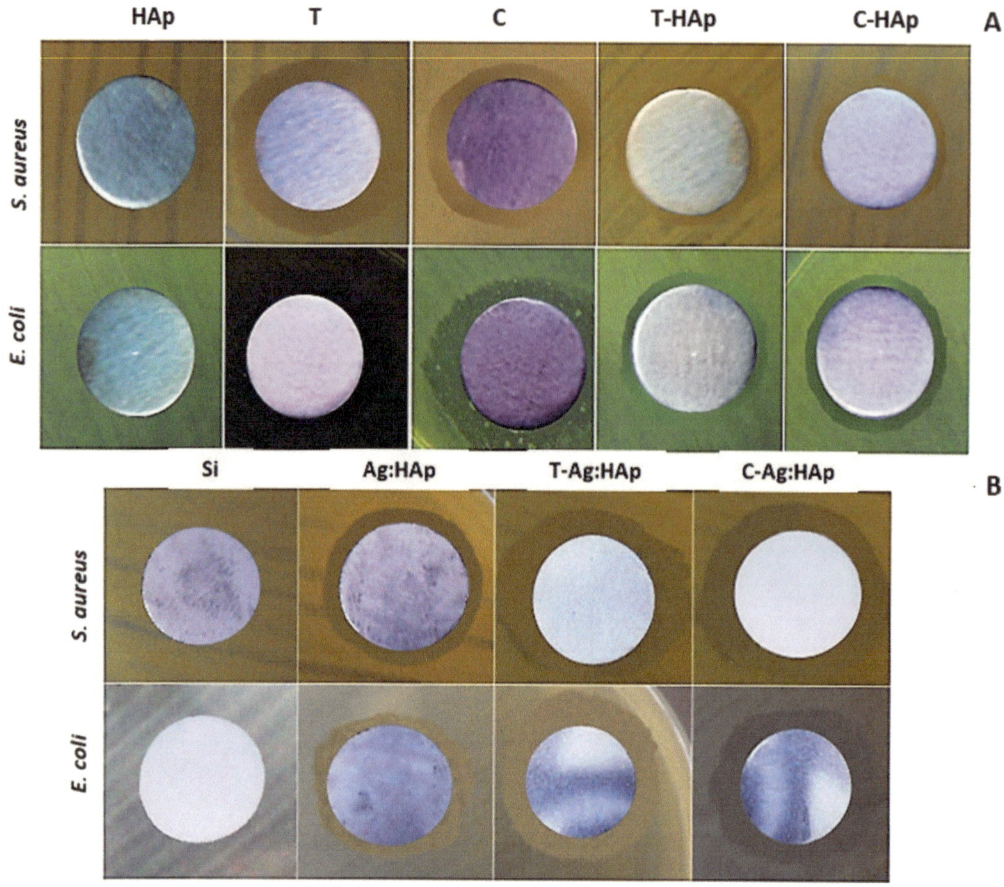

Figure 5. Antibacterial activity against *S. aureus* 0364 and *E. coli* ATCC 25922 cultures of (A) HAp, T-HAp and C-HAp, ciprofloxacin, tetracycline thin films and (B) Ag:HAp, T-Ag:HAp, C-Ag:HAp thin films. Reprinted from Ref. [125], the open access article distributed under the Creative Commons Attribution License.

Hydroxyapatite with an additional phase of sphere-shaped silver phosphate molecules and enriched with vancomycin or gentamicin has been developed by Suvannapruk et al. [126]. The authors have proved that such a combination of silver phosphate nanoparticles and antibiotic prolongs the antibacterial activity and increases the efficiency of the material.

An interesting experiment was proposed by Sampath Kumar [127], resulting in the creation of HA enriched with Ag^{+}, Sr^{2+} or Zn^{2+} ions. These materials were used as doxycycline-releasing media. Of all the materials under investigation, Ag-HA had the lowest doxycycline loading. The most optimal system was the Zn-HA, because it produced a sufficiently effective level of antibacterial activity and, at the same time, contained an adequate quantity of loaded antibiotic.

Recently, Yu et al. [128] synthesized Cu-HA microspheres using a microwave-hydrothermal method. Interestingly, the phosphorous source for the synthesis was creatine phosphate – a substrate for ATP production. Chitosan-based scaffolds were created by freeze drying and loaded with doxorubicin to examine drug loading and release. Osteogenic and angiogenic properties were evaluated both *in vitro* and *in vivo*. Samples were bioactive and non-toxic. The authors claim that the release of Cu^{2+} ions, by stabilizing HIF-1α, induced hypoxia in the bone tissue, which significantly stimulated neovascularization and improved bone regeneration.

5. Conclusions

Sophisticated porous hydroxyapatite structures and hydroxyapatite/polymer structures seem to offer potential as systems for the delivery of antibacterial agents directly into the bone. Thus, rather than delivering a single medicine, it would be possible to conduct combined therapy with various antibacterial agents with different dissolution profiles. The simultaneous application of antibiotics and HA modified by ions with antibacterial activity may contribute to development of the effective prevention and treatment methodology for post-surgical osseous inflammations. A therapy designed to directly target the affected area may significantly reduce general side effects of using antibiotics, improving therapeutic efficiency, while also allowing a reduction in dosage, which seems to be beneficial in both medical and economic terms.

Acknowledgements

This work was supported by the National Science Center (Poland) within project "Synthesis and physicochemical and biological analysis of crystalline calcium phosphates substituted with various ions"; UMO-2016/22/E/ST5/00564 and by Medical University of Warsaw (FW23/N/17).

Author details

Katarzyna Szurkowska, Aleksandra Laskus and Joanna Kolmas*

*Address all correspondence to: joanna.kolmas@wum.edu.pl

Faculty of Pharmacy with Laboratory Medicine Division, Department of Inorganic and Analytical Chemistry, Medical University of Warsaw, Warsaw, Poland

References

[1] Haider A, Haider S, Han SS, Kang I-K. Recent advances in the synthesis, functionalization and biomedical applications of hydroxyapatite: A review. RSC Advances. 2017;7:7442-7458. DOI: 10.1039/c6ra26124h

[2] Mucalo M, editor. Hydroxyapatite (Hap) for Biomedical Applications. 1st ed. Amsterdam: Elsevier; 2015. 404 p

[3] Marković S, Vaselinović L, Lukić MJ, Karanović L, Bracko I, Ignjatović N, Uskoković D. Synthetical bone-like and biological hydroxyapatites: A comparative study of crystal structure and morphology. Biomedical Materials. 2011;**6**:045005. DOI: org/10.1088/1748-6041/6/4/045005

[4] Dalmonico GML, Franczak PF, Levandowski N, Camargo NHA, Dallabrida AL, da Costa BD, Garcia Gil O, Cambra-Moo O, Rodriguez MA, Canillas M. An in vivo study on bone formation behaviour of microporous granular calcium phosphate. Biomaterials Science 2017;**5**:1315-1325. DOI: 10.1039/C7BM00162B

[5] Kasir R, Vernekar VN, Laurencin CT. Inductive biomaterials for bone regeneration. Materials Research. 2017;**32**:1047-1060. DOI: https://doi.org/10.1557/jmr.2017.39

[6] Koju N, Sikder P, Ren Y, Zhou H, Bhaduri SB. Biomimetic coating technology for orthopedic implants. Current Opinion in Chemical Engineering. 2017;**15**:49-55. DOI: https://doi.org/10.1016/j.coche.2016.11.005

[7] Elkassas D, Arafa A. The innovative applications of therapeutic nanostructures in dentistry. Nanomedicine: Nanotechnology, Biology and Medicine. 2017;**13**:1543-1562. DOI: https://doi.org/10.1016/j.nano.2017.01.018

[8] Ridi F, Meazzini I, Catroflorio B, Bonini M, Berti D, Baglioni P. Functional calcium phosphate composites in nanomedicine. Advance in Colloid and Interface Science. 2017;**244**:281-295. DOI: https://doi.org/10.1016/j.cis.2016.03.006

[9] Wang C, Wang Y, Meng H, Wang X, Zhu Y, Yu K, Yuan X, Wang A, Guo Q, Peng J. Research progress regarding nanohydroxyapatite and its composite biomaterials in bone defect repair. International Journal of Polymeric Materials and Polymeric Biomaterials. 2016; **65**:601-610

[10] Liu Q, Pan H, Chen Z, Matinlinna JP. Insight into bone-derived biological apatite: Ultrastructure and effect on thermal treatment. Biomed Research International. 2015;**2015**. Article ID: 601025. DOI: 10.1155/2015/601025

[11] Kolmas J, Velard F, Jaguszewska A, Lemaire F, Kerdjoudj H, Gangloff SC, Kaflak A. Substitution of strontium and boron into hydroxyapatite crystals: Effect on physicochemical properties and biocompatibility with human Wharton-Jelly stem cells. Materials Science and Engineering C. 2017;**79**:638-646. DOI: https://doi.org/10.1016/j.msec.2017.05.066

[12] Kolmas J, Pajor K, Pajchel L, Przekora A, Ginalska G, Oledzka E, Sobczak M. Fabrication and physicochemical characterization of porous composite microgranules with selenium oxyanions and risedronate sodium for potential applications in bone tumors. International Journal of Nanomedicine. 2017;**12**:1-10

[13] Lerner T, Liljenqvist U. Silicate-substituted calcium phosphate as a bone graft substitute in surgery for adolescent idiopathic scoliosis. European Spine Journal. 2013;**22**(S2): 185-194. DOI: https://doi.org/10.1007/s00586-012-2485-7

[14] Kolmas J, Krukowski S, Laskus A, Jurkitewicz M. Synthetic hydroxyapatite in pharmaceutical applications. Ceramics International. 2016;**42**:2472-2487. DOI: https://doi.org/10.1016/j.ceramint.2015.10.048

[15] Oledzka E, Sobczak M, Kolmas J, Nałęcz-Jawecki G. Selenium-substituted hydroxyapatite/biodegradable polimer/pamidronate combined scaffold for the therapy of bone tumor. International Journal of Molecular Sciences. 2015;**16**:22205-22222. DOI: 10.3390/ijms160922205

[16] De Miguel L, Popa I, Noiray M, Caudron E, Arpinati L, Desmaele D, Cebrian-Torrejon G, Domenech-Carbo A, Ponchel G. Osteotropic polypeptide nanoparticles with dual hydroxyapatite binding properties and controlled cisplatin delivery. Pharmaceutical Research. 2015;**32**:1794-1803. DOI: https://doi.org/10.1007/s11095-014-1576-z

[17] Uskoković V, Desai TA. Simultanous bactericidal and osteogenic effect of nanoparticulate calcium phosphate powders loaded with clindamycin on osteoblasts infected with *Staphylococcus aureus*. Materials Science and Engineering C. 2014;**37**:210-222. DOI: https://doi.org/10.1016/j.msec.2014.01.008

[18] von Stechov D, Rauschmann MA. Effectiveness of combination use of antibiotic-loaded PerOssal® with spinal surgery in patients with spondylodiscitis. European Surgical Research 2009;**43**:298-305. DOI: 10.1159/000233525

[19] Nandi SK, Mukherjee P, Roy S, Kundu B, Kumar De D, Basu D. Local antibiotic delivery systems for the treatment of osteomyelitis – A review. Materials Science and Engineering C. 2009;**29**:2478-2485. DOI: https://doi.org/10.1016/j.msec.2009.07.014

[20] Martínez-Vásquez F, Cabaños MV, Paris JL, Lozano D, Vallet-Regi M. Fabrication of novel Si-doped hydroxyapatite/gelatin scaffolds by rapid prototyping for drug delivery and bone regeneration. Acta Biomaterialia. 2015;**15**:200-209. DOI: 10.1016/j.actbio.2014.12.021

[21] Hasegawa M, Sudo A, Komlev VS, Barinov ML, Uchida A. High release of antibiotic from a novel hydroxyapatite with bimodal pore size distribution. Journal of Biomedical Materials Research B: Applied Biomaterials. 2004;**70**(2):332-339. DOI: 10.1002/jbm.b.30047

[22] Ferraz MP, Mateus AY, Sousa JC, Monteiro FJ. Nanohydroxyapatite microspheres as delivery system for antibiotics: Release kinetics, antimicrobial activity, and interaction with osteoblasts. Journal of Biomedical Materials Research A. 2007;**81**(4):994-1004. DOI: 10.1002/jbm.a.31151

[23] Sivakumar M, Rao KP. Preparation, characterization and in vitro release of gentamicin from coralline hydroxyapatite – alginate composite microspheres. Journal of Biomedical Materials Research A. 2003;**65**(2):222-228. DOI: 10.1002/jbm.a.10495

[24] Hess U, Mikolajczyk G, Treccani L, Streckbein P, Heiss C, Odenbach S, Rezwan K. Multi-loaded ceramic beads/matrix scaffolds obtained by combining ionotropic and freeze gelation for sustained and tuneable vancomycin release. Materials Science and Engineering C. 2016;**67**:542-553. DOI: 10.1016/j.msec.2016.05.042

[25] Yu J, Chu X, Cai Y, Ton P, Yao J. Preparation and characterization of antimicrobial nano-hydroxyapatite composites. Materials Science and Engineering C. 2014;**37**:54-59. DOI: 10.1016/j.msec.2013.12.038

[26] Thomas MB, Metoki N, Mandler D, Eliaz N. In situ, potentiostatic deposition of calcium phosphate with gentamicin-loaded chitosan nanoparticles on titanium alloy surfaces. Electrochimica Acta. 2016;**222**:355-360. DOI: https://doi.org/10.1016/j.electacta.2016.10.186

[27] Yang C-C, Lin C-C, Liao J-W, Yen S-K. Vancomycin-chitosan composite deposited on post porous hydroxyapatite coated Ti6Al4V implant for drug controlled release. Materials Science and Engineering C. 2013;**33**:2203-2212. DOI: https://doi.org/10.1016/j.msec.2013.01.038

[28] Teng SH, Lee EJ, Wang P, Yun SH, Han CM, Kim HE. Functionally gradient chitosan/hydroxyapatite composite scaffolds for controlled drug release. Journal of Biomedical Materials Research B. 2009;**90**(1):275-282. DOI: 10.1002/jbm.b.31283

[29] Ionita D, Bajenaru-Georgescu D, Totea G, Mazare A, Schmuki P, Demetrescu I. Activity of vancomycin release from bioinspired coatings of hydroxyapatite or TiO_2 nanotubes. International Journal of Pharmaceutics. 2017;**517**:296-302. DOI: 10.1016/j.ijpharm.2016.11.062

[30] Lian X, Liu H, Wang X, Xu S, Cui F, Bai X. Antibacterial and biocompatible properties of vancomycin-loaded nanohydroxyapatite/collagen/poly(lactic-acid) bone substitute. Progress in Natural Sciences: Materials International. 2013;**23**(6):549-556. DOI: https://doi.org/10.1016/j.pnsc.2013.11.003

[31] Lian X, Mao K, Liu X, Vang X, Cui F. In vivo osteogenesis of vancomycin loaded nano-hydroxyapatite/collagen/calcium sulphate composite for treating infectious bone defect induced by chronic osteomyelitis. Journal of Nanomaterials. 2015 Article ID: 261492. http://dx.doi.org/10.1155/2015/261492

[32] Song W, Ren W, Wan C, Esquivel AO, Shi T, Blasier R, Marker DC. A novel strontium-doped calcium polyphosphate/erythromycin/poly(vinyl alcohol) composite for bone tissue engineering. Journal of Biomedical Materials Research A. 2011;**98**(3):359-371. DOI: 10.1002/jbm.a.33127

[33] Sasikumar S. Effect of particle size of calcium phosphate based bioceramic drug delivery carrier on the release kinetics of ciprofloxacin hydrochloride: An in vitro study. Frontiers in Materials Science. 2013;**7**(3):261-268. DOI: 10.1007/s11706-013-0216-6

[34] Yu M, Zhou K, Li Z. Preparation, characterization and in vitro gentamicin release of porous HA microspheres. Materials Science and Engineering C. 2014;**45**:306-312. DOI: 10.1016/j.msec.2014.08.075

[35] Schnieders J, Gbureck U, Thull R, Kissel T. Controlled release of gentamicin from calcium phosphate-poly(lactic acid-co-glycolic acid) composite bone cement. Biomaterials. 2006;**7**:4239-4249. DOI: 10.1016/j.biomaterials.2006.03.032

[36] Wang X, Xu H, Zhao Y. Poly(lactide-co-glycolide) encapsulated hydroxyapatite microspheres for sustained release of doxycycline. Materials Science and Engineering B. 2012;**177**:367-372. DOI: https://doi.org/10.1016/j.mseb.2011.12.030

[37] Tang Y, Chen L, Zhao K, Wu Z, Wang Y, Tan Q. Fabrication of PLGA/HA (core)-collagen/amoxicillin (shell) nanofiber membranes through coaxial electrospinning for guided tissue regeneration. Composites Science and Technology. 2016;**125**:100-107

[38] Zheng F, Wang S, Wen S, Shen M, Zhu M, Shi X. Characterization and antibacterial activity of amoxicillin-loaded electrospunnano-hydroxyapatite/poly(lactic-co-glycolic acid) composite nanofibers. Biomaterials. 2013;**34**:1402-1412. DOI: 10.1016/j.biomaterials.2012.10.071

[39] Vukomanovic M, Skapin SD, Poljansec I, Zagar E, Kralj B, Ignjatovic N, Uskoković D. Poly(D,L-lactide-co-glycolide)/hydroxyapatite core-shell nanosphere. Part 2: Simultaneous release of a drug and a prodrug (clindamycin and clindamycin phosphate). Colloids and Surface B: Biointerfaces. 2011;**82**(2):414-421. DOI: https://doi.org/10.1016/j.colsurfb.2010.09.012

[40] Uskokovic V, Hoover C, Vukomanovic M, Uskoković DP, Desai TA. Osteogenic and antimicrobial nanoparticulate calcium phosphate and poly-(D,L-lactide-co-glycolide) powders for the treatment of osteomyelitis. Materials Science and Engineering C. 2013;**33**:3362-3373. DOI: 10.1016/j.msec.2013.04.023

[41] Yan L, Jiang DM, Cao Z-D, Wu J, Wang X, Wang ZL, Li YJ, Yi YF. Treatment of *Staphylococcus aureus*-induced chronic osteomyelitis with bone-like hydroxyapatite/poly amino acid loaded with rifapentine microspheres. Drug Design, Development and Therapy. 2015;**9**:3665-3676. DOI: https://doi.org/10.2147/DDDT.S84486

[42] Leprêtre S, Chai F, Hornez J-C, Vermet G, Neut C, Descamps M, Hildebrand HF, Martel B. Prolonged local antibiotics delivery from hydroxyapatite functionalized with cyclodextrin polymers. Biomaterials. 2009;**30**:6086-6093. DOI: 10.1016/j.biomaterials.2009.07.045

[43] Selvakumar M, Kumar PS, Das B, Dhara S, Chattopadhyay S. Structurally tuned antimicrobial mesoporous hydroxyapatite nanorods by cyclic oligosaccharides regulation to release a drug for osteomyelitis. Crystal Growth and Design. 2017;**17**:433-445. DOI: 10.1021/acs.cgd.6b01190

[44] Stigter M, Bezemer J, de Groot K, Layrolle P. Incorporation of different antibiotics into carbonated hydroxyapatite coatings on titanium implants, release and antibiotic efficacy. Journal of Controlled Release. 2004;**99**(1):127-137

[45] Mir M, Siddiqi SA, Hussain T, et al. Synthesis and characterization of calcium deficient apatite granules for drug eluting bone graft applications. Ceramics International. 2014;**40**:10719-10725. DOI: https://doi.org/10.1016/j.ceramint.2014.03.059

[46] Brohede U, Forsgren J, Roos S, Mihranyan A, Engqvist H, Stromme M. Multifunctional implant coatings providing possibilities for fast antibiotics loading with subsequent slow release. Journal of Materials Science: Materials in Medicine. 2009;**20**:1859-1867. DOI: 10.1007/s10856-009-3749-6

[47] Bhattacharya R, Kundu B, Nandi SK, Basu D. Systematic approach to treat chronic osteomyelitis through localized drug delivery system: Bench to bed side. Materials Science and Engineering C. 2013;**33**:3986-3993. DOI: 10.1016/j.msec.2013.05.036

[48] Chai F, Hornez J-C, Blanchemain N, Neut C, Descamps M, Hildebrandt HF. Antibacterial activation of hydroxyapatite (HA) with controlled porosity by different antibiotics. Biomolecular Engineering. 2007;**24**(5):510-514. DOI: 10.1016/j.bioeng.2007.08.001

[49] Ghosh S, Wu V, Pernal S, Uskoković V. Self-setting calcium phosphate cements with tunable antibiotic release rates for advanced antimicrobial applications. ACS Applied Materials Interfaces. 2016;**8**:7691-7708. DOI: 10.1021/acsami.6b01160

[50] Uskokovic V, Desai TA. Simultaneous bactericidal and osteogenic effect of nanoparticulate calcium phosphate powders loaded with clindamycin on osteoblasts infected with Staphylococcus aureus. Materials Science and Engineering C. 2014;**37**:210-222. DOI: 10.1016/j.msec.2014.01.008

[51] Victor SP, Sharma CP, Sreenivasan K. Use of quartz crystal nanobalance to study the binding and stabilization of albumin and doxycycline on a thin layer of hydroxyapatite. Applied Surface Science. 2011;**258**:1666-1669. DOI: https://doi.org/10.1016/j.apsusc.2011.09.119

[52] Canal C, Pastorino D, Mestres G, Schuler P, Ginebra MP. Relevance of microstructure for the early antibiotic release of fresh and pre-set calcium phosphate cements. Acta Biomaterialia. 2013;**9**:8403-8412

[53] Gomes PS, Santos JD, Fernandes MH. Cell-induced response by tetracyclines on human bonemarrow colonized hydroxyapatite and Bonelike®. Acta Biomaterialia. 2008;**4**:630-637

[54] Lilja M, Sörensen JH, Brohede U, Astrand M, Procter P, Arnoldi J, Steckel H, Stromme M. Drug loading and release of tobramycin from hydroxyapatite coated fixation pins. Journal of Materials Science: Materials in Medicine. 2013;**24**:2265-2274. DOI: 10.1007/s10856-013-4979-1

[55] Kaya M, Simsek-Kaya G. Gürsan N, Girecci E, Dayi E, Gundogdu B. Local treatment of chronic osteomyelitis with surgical debridement and tigecycline-impregnated calcium hydroxyapatite: An experimental study Oral Surgery Oral Medicine Oral Pathology Oral Radiology. 2012;**113**(3):340-347. DOI:10.1016/j.tripleo.2011.03.032

[56] Colovic A, Pasalic S, Jokanovic V. Influence of hydroxyapatite pore geometry on tigecycline release kinetics. Ceramics International. 2012;**38**:6181-6189. DOI: https://doi.org/10.1016/j.ceramint.2012.04.069

[57] Guo Y-J, Long T, Chen W, Ning C-Q, Zhu Z-A, Guo Y-P. Bactericidal property and biocompatibility of gentamicin-loaded mesoporous carbonated hydroxyapatite microspheres. Materials Science and Engineering C. 2013;**33**:3583-3591. DOI: https://doi.org/10.1016/j.msec.2013.04.021

[58] Joosten U, Joist A, Frebel T, Brandt B, Diederichs S, von Eiff C. Evaluation of an in situ setting injectable calcium phosphate as a new carrier material for gentamicin in the treatment of chronic osteomyelitis: Studies in vitro and in vivo. Biomaterials 2004;**25**:4287-4295. https://doi.org/10.1016/j.biomaterials.2003.10.083

[59] Alt V, Bitschnau A, Österling J, Sewing A, Meyer C, Kraus R, Meissner SA, Wenisch S, Domann E, Schnettler R. The effects of combined gentamicin–hydroxyapatite coating for cementless joint prostheses on the reduction of infection rates in a rabbit infection prophylaxis model. Biomaterials. 2006;**27**:4627-4634. DOI: 10.1016/j.biomaterials.2006.04.035

[60] Thanyaphoo S, Kaewsrichan J. Potential of bone scaffolds containing vancomycin and bone morphogenetic protein-2 in a rat model of osteomyelitis. Asian Biomedicine. 2014;**8**(5):651-657

[61] Joosten U, Joist A, Gosheder G, Liljenqvist U, Brandt B, von Eiff C. Effectiveness of hydroxyapatite-vancomycin bone cement in the treatment of *Staphylococcus aureus* induced chronic osteomyelitis. Biomaterials 2005;**26**:5251-5258. https://doi.org/10.1016/j.biomaterials.2005.01.001

[62] Jiang P–J, Patel S, Gbureck U, et al. Comparing the efficacy of three bioceramic matrices for the release of vancomycin hydrochloride. Journal of Biomedical Materials Research B Applied Biomaterials. 2010;**93**(1):58-61

[63] Guo Y-P, Yao Y-B, Guo Y-J, Ning C-Q. Hydrothermal fabrication of mesoporous carbonated hydroxyapatite microspheres for a drug delivery system. Microporous and Mesoporous Materials. 2012;**155**:245-251. DOI: https://doi.org/10.1016/j.micromeso.2012.01.037

[64] Perez LM, Lalueza P, Monzon M, Puertolas JA, Arruebo M, Santamaria J. Hollow porous implants filled with mesoporous silica particles as a two-stage antibiotic-eluting device. International Journal of Pharmaceutics. 2011;**409**:1-8. DOI: 10.1016/j.ijpharm.2011.02.015

[65] Rauschmann MA, Wichelhaus TA, Stirnal V, Dingeldein E, Zichner L, Schnettler R, Alt V. Nanocrystalline hydroxyapatite and calcium sulphate as biodegradable composite carrier material for local delivery of antibiotics in bone infections. Biomaterials. 2005;**26**:2677-2684. DOI: 10.1016/j.biomaterials.2004.06.045

[66] Belcarz A, Ginalska G, Zalewska J, Rzeski W, Slósarczyk A, Kowalczuk D, Godlewski P, Niedźwiadek J. Covalent coating of hydroxyapatite by keratin stabilizes gentamicin release. Journal of Biomedical Materials Research B Applied Biomaterials. 2009;**89**(1): 102-113. DOI: 10.1002/jbm.b.31192

[67] Garner S, Barbour ME. Nanoparticles for controlled delivery and sustained release of chlorhexidine in the oral environment. Oral Diseases. 2015;**21**(5):641-644. DOI: 10.1111/odi.12328

[68] Lara HH, Garza-Trevino EN, Ixtepan-Turrent L, Singh DK. Silver nanoparticles are broad-spectrum bactericidal and virucidal compounds. Journal of Nanobiotechnology. 2011;**9**:30. DOI: 10.1186/1477-3155-9-30

[69] Rai MK, Deshmukh SD, Ingle AP, Gade AK. Silver nanoparticles: The powerful nanoweapon against multidrug-resistant bacteria. Journal of Applied Microbiology. 2012;**112**:841-852. DOI: 10.1111/j.1365-2672.2012.05253.x

[70] Akiyama T, Miyamoto H, Yonekura Y, Tsukamoto M, Ando Y, Noda I, Sonohata M, Mawatari *M. silver* oxide-containing hydroxyapatite coatings has in vivo antibacterial activity in the rat tibia. Journal of Orthopaedic Research. 2013;**31**:1195-1200. DOI: 10.1002/jor.22357

[71] Anjaneyulu U, Priyadarshini B, Nirmala Grace A, Vijayalakshmi U. Fabrication and characterization of Ag doped hydroxyapatite-polyvinyl alcohol composite nanofibers and its in vitro biological evaluations for bone tissue engineering applications. Journal of Sol-Gel Science and Technology. 2017;**81**:750-761. DOI: 10.1007/s10971-016-4243-5

[72] Ipekoglu M, Altintas S. Silver substituted nanosized calcium deficient hydroxyapatite. Materials Technology. 2010;**25**:295-301. DOI: 10.1179/175355510X12692596613648

[73] Ciuca S, Badea M, Pozna E, Pana I, Kiss A, Floroian L, Semenescu A, Cotrut CM, Moga M, Vladescu A. Evaluation of Ag containing hydroxyapatite coatings to the *Candida albicans*infection. Journal of Microbiological Methods. 2016;**125**:12-18. DOI: 10.1016/j.mimet.2016.03.016

[74] Ciobanu CS, Iconaru SL, Coustumer PL, Constantin LV, Predoi D. Antibacterial activity of silver-doped hydroxyapatite nanoparticles against gram-positive and gram-negative bacteria. Nanoscale Research Letters. 2012;**7**:324

[75] Iconaru SL, Chifiriuc MC, Groza A. Structural and antimicrobial evaluation of silver doped hydroxyapatite-polydimethylsiloxane thin layers. Journal of Nanomaterials. 2017;**2017**:7492515. DOI: 10.1155/2017/7492515

[76] Miranda M, Fernández A, Díaz M, Esteban-Tejeda L, López-Esteban S, Malpartida F, Torrecillas R, Moya JS. Silver-hydroxyapatite nanocomposites as bactericidal and fungicidal materials. International Journal of Materials Research. 2010;**101**:122-127. DOI: 10.3139/146.110256

[77] Stanić V, Janaćković D, Dimitrijević S, Tanasković SB, Mitrić M, Pavlović MS, Krstić A, Jovanović D, Raičević S. Synthesis of antimicrobial monophase silver-doped hydroxyapatite nanopowders for bone tissue engineering. Applied Surface Science. 2011;**257**:4510-4518. DOI: 10.1016/j.apsusc.2010.12.113

[78] Jegatheeswaran S, Sundrarajan M. PEGylation of novel hydroxyapatite/PEG/Ag nanocomposite particles to improve its antibacterial efficacy. Material Science and Engineering C. 2015;**51**:174-181. DOI: 10.1016/j.msec.2015.02.012

[79] Shi C, Gao J, Wang M, Fu J, Wang D, Zhu Y. Ultra-trace silver-doped hydroxyapatite with non-cytotoxicity and effective antibacterial activity. Materials Science and Engineering C. 2015;**55**:497-505. DOI: 10.1016/j.msec.2015.05.078

[80] Yan Y, Zhang X, Huang Y, Ding Q, Pang X. Antibacterial and bioactivity of silver substituted hydroxyapatite/TiO_2 nanotube composite coatings on titanium. Applied Surface Science. 2014;**314**:348-357. DOI: 10.1016/j.apsusc.2014.07.027

[81] Popa CL, Ciobanu CS, Voicu G, Vasile E, Chifiriuc MC, Iconaru SL, Predoi D. Influence of thermal treatment on the antimicrobial activity of silver-doped biological apatite. Nanoscale Research Letters. 2015;**10**:502. DOI: 10.1186/s11671-015-1211-x

[82] Nath S, Kalmodia S, Basu B. Densification, phase stability and in vitro biocompatibility property of hydroxyapatite-10 wt% silver composites. Journal of Materials Science: Materials in Medicine. 2010;**21**:1273-1287. DOI: 10.1007/s10856-009-3939-2

[83] Rajendran A, Barik RC, Natarajan D, Kiran MS, Pattanayak DK. Synthesis, phase stability of hydroxyapatite-silver composite with antimicrobial activity and cytocompatability. Ceramics International. 2014;**40**:10831-10838. DOI: 10.1016/j.ceramint.2014.03.075

[84] Liu H, Man HC. Laser fabrication of Ag-HA nanocomposites on Ti6Al4V implant for enhancing bioactivity and antibacterial capability. Materials Science and Engineering C. 2017;**70**:1-8. DOI: 10.1016/j.msec.2016.08.059

[85] Lu X, Zhang B, Wang Y, Zhou X, Weng J, Qu S, Feng B, Watari F, Ding Y, Leng Y. Nano-Ag-loaded hydroxyapatite coatings on titanium surfaces by electrochemical deposition. Journal of Royal Society Interface. 2011;**8**:529-539. DOI: 10.1098/rsif.2010.0366

[86] Lee DH, Min BG. Preparation and antibacterial properties of nanocomposite fibers made of polyamide 6 and silver-doped hydroxyapatite. Fibers and Polymers. 2014;**15**: 1921-1926. DOI: 10.1007/s12221-014-1921-1

[87] Xie CM, Lu X, Wang KF, Meng FZ, Jiang O, Zhang HP, Zhi W, Fang LM. Silver nanoparticles and growth factors incorporated hydroxyapatite coatings on metallic implant surfaces for enhancement of osteoinductivity and antibacterial properties. ACS Applied Materials Interfaces. 2014;**6**:8580-8589. DOI: 10.1021/am501428e

[88] Lu M, Liao J, Dong J, Wu J, Qiu H, Zhou X, Li J, Jiang D, He TC, Quan Z. An effective treatment of experimental osteomyelitis using the antimicrobial titanium/silver-containing nHP66 (nano-hydroxyapatite/polyamide-66) nanoscaffold biomaterials. Scientific Reports. 2016;**6**:39174. DOI: 10.1038/srep39174

[89] Ueno M, Miyamoto H, Tsukamoto M, Eto S, Noda I, Shobuike T, Kobatake T, Sonohata M, Mawatari M. Silver-containing hydroxyapatite coating reduces biofilm formation by methicillin-resistant *Staphylococcus aureus* in vitro and in vivo. BioMed Research International. 2016;**2016**:8070597. DOI: 10.1155/2016/8070597

[90] Eto S, Kawano S, Someya S, Miyamoto H, Sonohata M, Mawatari M. First clinical experience with thermal-sprayed silver oxide-containing hydroxyapatite coating implant. The Journal of Arthroplasty. 2016;**31**:1498-1503. DOI: 10.1016/j.arth.2015.12.034

[91] Lim PN, Shi Z, Neoh KG, Ho B, Tay BY, Thian ES. The effects of silver, silicon-containing apatite towards bacteria and cell responses. Biomedical Materials. 2014;**9**:015010. DOI: 10.1088/1748-6041/9/1/015010

[92] Xu Z, Lei Y, Yin W, Chen Y, Ke Q, Guo Y, Zhang C. Enhanced antibacterial activity and osteoinductivity of Ag-loaded strontium hydroxyapatite/chitosan porous scaffold for bone tissue engineering. Journal of Materials Chemistry B. 2016;**4**:7919-7928. DOI: 10.1039/C6TB01282E

[93] Aksakal B, Demirel M. The effect of Zirconia/Yttria/Silver substitutions on mechanostructure and cell viability of the synthesized bioceramic bone grafts. Ceramics International. 2017;**43**:7482-7487. DOI: 10.1016/j.ceramint.2017.03.026

[94] Kolmas J, Piotrowska U, Kuras M, Kurek E. Effect of carbonate substitution on physicochemical and biological properties of silver containing hydroxyapatites. Materials Science and Engineering C. 2017;**74**:124-130. DOI: 10.1016/j.msec.2017.01.003

[95] Huo K, Zhang X, Wang H, Zhao L, Liu X, Chu PK. Osteogenic activity and antibacterial effects on titanium surfaces modified with Zn-incorporated nanotube arrays. Biomaterials. 2013;**34**:3467-3478. DOI: 10.1016/j.biomaterials.2013.01.071

[96] Cai R, Wang H, Cao M, Hao L, Zhai L, Jiang S, Li X. Synthesis and antimicrobial activity of mesoporous hydroxylapatite/zinc oxide nanofibers. Materials and Design. 2015;**87**:17-24. DOI: 10.1016/j.matdes.2015.08.004

[97] Fang J, Zhao J, Sun Y, Ma H, Yu X, Ma Y, Ni Y, Zheng L, Zhou Y. Biocompatibility and antibacterial properties of zinc-ion implantation on titanium. Journal of Hard Tissue Biology. 2014;**23**:35-44

[98] Anwar A, Akbar S, Sadiqa A, Kazmi M. Novel continuous flow synthesis, characterization and antibacterial studies of nanoscale zinc substituted hydroxyapatite bioceramics. Inorganica Chimica Acta. 2016;**453**:16-22. DOI: 10.1016/j.ica.2016.07.041

[99] Chen X, Tang QL, Zhu YJ, Zhu CL, Feng XP. Synthesis and antibacterial property of zinc loaded hydroxyapatite nanorods. Materials Letters. 2012;**89**:233-235. DOI: 10.1016/j.matlet.2012.08.115

[100] Predoi D, Iconaru SL, Deniaud A, Chevallet M, Michaud-Soret I, Buton N, Prodan AM. Textural, structural and biological evaluation of hydroxyapatite doped with zinc at low concentrations. Materials. 2017;**10**:229. DOI: 10.3390/ma10030229

[101] Radovanović Ž, Veljović D, Jokić B, Dimitrijević S, Bogdanović G, Kojić V, Petrović R, Janaćković D. Biocompatibility and antimicrobial activity of zinc(II)-doped hydroxyapatite, synthesized by a hydrothermal method. Journal of Serbian Chemical Society. 2012;**77**:1787-1798. DOI: 10.2298/JSC121019131R

[102] Thian ES, Konishi T, Kawanobe Y, Lim PN, Choong C, Aizawa M. Zinc-substituted hydroxyapatite: A biomaterial with enhanced bioactivity and antimicrobial properties. Journal of Materials Science: Materials in Medicine. 2013;**24**:437-445. DOI: 10.1007/s10856-012-4817-x

[103] Tank KP, Chudasama KS, Thaker VS, Joshi MJ. Pure and zinc doped nano-hydroxyapatite: Synthesis, characterization, antimicrobial and hemolytic studies. Journal of Crystal Growth. 2014;**401**:474-479. DOI: 10.1016/j.jcrysgro.2014.01.062

[104] Zhang J. Biocompatibility and anti-bacterial activity of Zn-containing HA/TiO_2 hybrid coatings on Ti substrate. Journal of Hard Tissue Biology. 2013;**22**:311-318

[105] Colombo M, Beltrami R, Rattalino D, Mirando M, Chiesa M, Poggio C. Protective effects of a zinc-hydroxyapatite toothpaste on enamel erosion. SEM study. Annali di Stomatologia (Roma). 2017;**10**:38-45. DOI: 10.11138/ads/2016.7.3.038

[106] Samani S, Hossainalipour SM, Tamizifar M, Rezaie HR. *In vitro* antibacterial evaluation of sol-gel-derived Zn-, Ag-, and (Zn + Ag)-doped hydroxyapatite coatings against methicillin-resistant *Staphylococcus aureus*. Journal of Biomedical Materials Research Part A. 2013;**101A**:222-230. DOI: 10.1002/jbm.a.34322

[107] Livitska O, Strutynska N, Zatovsky I, Nikolenko I, Slobodyanik N, Prytutskyy Y, Epple M, Prymak O, Byeda A. Copper (II), zinc (II) and copper (II)/zinc (II)-containing carbonate-substituted hydroxyapatite: synthesis, characterization and thermal behavior. Materialwissenschaft und Werkstofftechnik. 2016;**47**:85-91. DOI: 10.1002/mawe.2016 00460

[108] Kaygili O, Keser S. Sol-gel synthesis and characterization of Sr/Mg, Mg/Zn and Sr/Zn co-doped hydroxyapatites. Materials Letters. 2015;**141**:161-164. DOI: 10.1016/j.matlet.2014.11.078

[109] Friederichs RJ, Chappell HF, Shepherd DV, Best SM. Synthesis, characterization and modelling of zinc and silicate co-substituted hydroxyapatite. Journal of Royal Society Interface. 2015;**12**:20150190. DOI: 10.1098/rsif.2015.0190

[110] Uysal I, Severcan F, Evis Z. Characterization by Fourier transform infrared spectroscopy of hydroxyapatite co-doped with zinc and fluoride. Ceramics International. 2013;**39**:7727-7733. DOI: 10.1016/j.ceramint.2013.03.029

[111] Vincent M, Hartemann P, Engels-Deutsch M. Antimicrobial applications of copper. International Journal of Hygiene and Environmental Health. 2016;**219**:585-591. DOI: 10.1016/j.ijheh.2016.06.003

[112] Li Y, Ho J, Ooi CP. Antibacterial efficacy and cytotoxicity studies of copper (II) and titanium (IV) substituted hydroxyapatite nanoparticles. Materials Science and Engineering C. 2010;**30**:1137-1144. DOI: 10.1016/j.msec.2010.06.011

[113] Sahithi K, Swetha M, Prabaharan M, Moorthi A, Saranya N, Ramasamy K, Srinivasan N, Partridge NC, Selvamurugan N. Synthesis and characterization of nanoscale-hydroxyapatite-copper for antimicrobial activity towards bone tissue engineering applications. Journal of Biomedical Nanotechnology. 2010;**6**:333-339. DOI: 10.1166/jbn.2010.1138

[114] Shanmugam S, Gopal B. Copper substituted hydroxyapatite and fluorapatite: Synthesis, characterization and antimicrobial properties. Ceramics International. 2014;**40**:15655-15662. DOI: 10.1016/j.ceramint.2014.07.086

[115] Stanić V, Dimitrijević S, Antić-Stanković J, Mitrić M, Jokić B, Plećaš IB, Raičević S. Synthesis, characterization and antimicrobial activity of copper and zinc-doped hydroxyapatite nanopowders. Applied Surface Science. 2010;**256**:6083-6089. DOI: 10.10 16/j.apsusc.2010.03.124

[116] Radovanović Ž, Jokić B, Veljović D, Dimitrijević S, Kojić V, Petrović R, Janaćković D. Antimicrobial activity and biocompatibility of Ag^{+}- and Cu^{2+}-doped biphasic hydroxyapatite/α-tricalcium phosphate obtained from hydrothermally synthesized Ag^{+}- and Cu^{2+}-doped hydroxyapatite. Applied Surface Science. 2014;**307**:513-519. DOI: 10.1016/j.apsusc.2014.04.066

[117] Tran PL, Hammond AA, Mosley T, Cortez J, Gray T, Colmer-Hamood JA, Shashtri M, Spallholz JE, Hamood AN, Reid TW. Organoselenium coating on cellulose inhibits the formation of biofilms by *Pseudomonas aeruginosa* and *Staphylococcus aureus*. Applied Environmental Microbiology. 2009;**75**:3586-3592. DOI: 10.1128/AEM.02683-08

[118] Tran PA, Webster TJ. Selenium nanoparticles inhibit *Staphylococcus aureus* growth. International Journal of Nanomedicine. 2011;**6**:1553-1558. DOI: 10.2147/IJN.S21729

[119] Rodríguez-Valencia C, López-Álvarez M, Cochón-Cores B, Pereiro I, Serra J, González P. Novel selenium-doped hydroxyapatite coatings for biomedical applications. Journal of Biomedical Materials Research Part A. 2013;**101**:853-861. DOI: 10.1002/jbm.a.34387

[120] Uskoković V, Iyer MA, Wu VM. One ion to rule them all: the combined antibacterial, osteoinductive and anticancer properties of selenite-incorporated hydroxyapatite. Journal of Materials Chemistry B. 2017;**5**:1430-1445. DOI: 10.1039/C6TB03387C

[121] Kolmas J, Groszyk E, Piotrowska U. Nanocrystalline hydroxyapatite enriched in selenite and manganese ions: Physicochemical and antibacterial properties. Nanoscale Research Letters. 2015;**10**:278. DOI: 10.1186/s11671-015-0989-x

[122] Kolmas J, Groszyk E, Kwiatkowska-Różycka D. Substituted hydroxyapatites with antibacterial properties. BioMed Research International. 2014;**2014**:178123. DOI: 10.1155/2014/178123

[123] Ivashenko OA, Perekos AO, Ulianchych NV, Uvarova IV, Protsenko LS, Budylina OM, Holovkova MY, Yarmola TM. Interaction of Ag-free and Ag-doped hydroxyapatite with ciprofloxacin solutions. Materialwissenschaft und Werkstofftechnik. 2011;**42**: 98-108. DOI: 10.1002/mawe.201100739

[124] Xiong Z-C, Yang Z-Y, Zhu Y-J, Chen F-F, Zhang Y-G, Yang R-L. Ultralong hydroxyapatite nanowires-based paper co-loaded with silver nanoparticles and antibiotic for long-term antibacterial benefit. ACS Applied Materials and Interfaces. 2017;**9**(27):22212-22222. DOI: 10.1021/acsami.7b05208

[125] Predoi D, Popa CL, Chapon P, Groza A, Iconaru SL. Evaluation of the antimicrobial activity of different antibiotics enhanced with silver-doped hydroxyapatite thin-films. Materials. 2016;**9**(9),778:1-18. DOI: 10.3390/ma9090778

[126] Suvannapruk W, Thammarakcharoen F, Phanpiriya P, Suwanprateeb J. Development of antibiotic impregnated nanosized silver phosphate-doped hydroxyapatite bone-graft. Journal of Nanomaterials. 2013, Article ID:542584:1-9. http://dx.doi.org/10.1155/2013/542584

[127] Sampath Kumar TS, Madhumathi K, Rubalya Y, Doble M. Dual mode antibacterial activity of ion substituted calcium phosphate nanocarriers for bone infections. Frontiers in Bioengineering and Biotechnology. 2015;**3**(59):1-10. DOI: 10.3389/fbioe.2015.00059

[128] Yu W, Sun T, Ding Z, Qi C, Zhao H, Chen F, Shi Z, Zhu Y, Chen D, He J. Copper-doped mesoporous hydroxyapatite microspheres synthesized by a microwave-hydrothermal method using creatine phosphate as an organic phosphorus source: application in drug delivery and enhanced bone regeneration. Journal of Materials Chemistry B. 2017;**5**:1039-1052. DOI: 10.1039/C6TB02747D

Chapter 8

316L Stainless Steel/Hydroxyapatite Composite Materials for Biomedical Applications

Aurora Anca Poinescu and Rodica-Mariana Ion

Additional information is available at the end of the chapter

http://dx.doi.org/10.5772/intechopen.71490

Abstract

Hydroxyapatite (HAp) is known as one of the most important biomaterials used for orthopedic and dental applications due to chemical composition similar to that of bone and bioactive properties. However, due to its reduced resistance, HAp should be mixed with different components in order to create composite materials. A good example is austenitic stainless steel (AA 316L), as a class of metallic materials used for biomedical applications. The aim of this chapter is to show a series of our group studies for obtaining a hybrid metal-ceramic composite by powder metallurgy, the physicochemical, mechanical, and biological properties of these hybrid composites and the investigation of possible functional coating layers and adherent bioactive hydroxyapatite $Ca_{10}(PO_4)_6(OH)_2$ on the metal of 316L stainless steel, correlating our data with literature data. Hardness properties of the obtained composites are close to those of human bones. In biological systems, could be observed that after 6 months, the relevant metals (Fe, Mn, Cr, and Ni) concentration (ppb), that may be released from composite materials in simulated physiological fluids (SBF), is practically zero.

Keywords: hydroxyapatite (HAp), nanotechnology, 316L, HAp thin films

1. Introduction

A definition of "biomaterial," approved by a consensus of experts, is a nonviable material used in a device, which could interact with biological systems [1]. Except biological area, this definition covers a wide range of applications. Biocompatibility is a complex concept that considers all processes that occur in the interaction between biomaterial and a living organism. The human body is highly aggressive in terms of corrosion for proper metallic biomaterials used in the manufacture of implants. This is the reason of intensive selection of new and proper biomaterials [2].

In recent decades, several strategies have been developed to obtain biomaterials used for medical implants. The development of this field is necessary for obtaining materials with improved characteristics. Hydroxyapatite is one of the most important biomaterials used for orthopedic and dental applications due to its chemical composition like that of bone and bioactive properties [3, 4]. Austenitic stainless steels are a class of metallic materials used for biomedical applications, too. Also, 316L stainless steel has a high corrosion resistance, good mechanical properties, and relatively low cost of obtaining and processing [5].

Experimental researches in this chapter will include the following groups of materials and methods:

- Ceramics: hydroxyapatite.
- Metallic materials: 316L stainless steel.
- Composite materials HAp/316L obtained by powder metallurgy.
- Composite materials HAp/316L thin films obtained by sol-gel.
- Functional coating layers and adherent bioactive hydroxyapatite $Ca_{10}(PO_4)_6(OH)_2$ on the metal of 316L stainless steel.

In this chapter the following aspects will be discussed: (i) synthesis of HAp by wet precipitation; (ii) synthesis of hydroxyapatite by sol-gel method; (iii) chemical, mechanical, and microstructural characterization of 316L stainless steel and HAp (X-ray diffraction (XRD), dynamic light scattering (DLS), Fourier transform infrared spectroscopy (FTIR), scanning electron microscopy (SEM), atomic force microscopy (AFM), mechanical (the tensile behavior, Vickers hardness, surface roughness)); (iv) synthesis and characterization of 316L steel and hydroxyapatite composites by powder metallurgy, sol-gel, and functional coating layers between both components; and (v) study of the new biomaterials behavior in simulated physiological fluids, considering that the deposition of hydroxyapatite will prevent the release of metal ions in solutions of artificial plasma and saline.

2. State of research in biomaterials

2.1. Preparation of hydroxyapatite by wet chemical precipitation method

This method is easy to achieve, at low temperature, with relatively high percentage of pure product and with not very expensive equipments [6, 7].

Hydroxyapatite could be obtained by chemical precipitation method modified by Sung [8]. Commercial chemical reagents used are calcium nitrate tetrahydrate $Ca(NO_3)_2.4H_2O$ and ammonium dihydrogen phosphate $(NH_4)_2HPO_4$, separately dissolved in distilled water with stirring. After addition of $Ca(NO_3)_2$ in aqueous solution over the $(NH_4)_2HPO_4$, and shaken at room temperature for about 1 h, a milky precipitate, somewhat gelatinous, has been obtained which in turn was stirred for 1 h for a perfect mixing [9, 10]. The mixture thus obtained was sintered at 100°C for 24 h. After that the precipitate was washed and filtered on a glass filter. After filtration, sticky compacted product was dried in an oven at 80°C. Then, dry powder was crushed in a mortar and pestle and then calcinated in a crucible of alumina at three different temperatures: 800, 1000, and 1200°C for 1 h.

The best results have been obtained at a temperature of 1000°C, at the same molar concentrations and the same method of refluxing.

The samples calcinated at different temperatures (850, 1000, and 1200°C) were analyzed by spectral techniques: X-ray diffraction (XRD), Fourier transform infrared spectroscopy (FTIR), Raman spectroscopy, dynamic light scattering (DLS), scanning electron microscopy with energy-dispersive spectroscopy (SEM-EDS), and atomic force microscopy (AFM).

The powder synthesized chemically is consistent with diffraction reference ASTM pure HAp [11, 12]. Identification of phases identified in the synthesized hydroxyapatite allow the network parameters calculation by setting 2θ. For this purpose, it has been taken into account the width of diffraction peaks at half height of them, the lines having hkl indices (200), (002), (102), (210),(310), and (004), in accordance with literature data [13, 14].

The most intense peaks appear in 2θ between 20° and 60°. X-ray diffraction data by peak characteristic (002), (211), (112), and (300) showed a high degree of purity of hydroxyapatite (**Figure 1**). This is confirmed by X-ray diffraction that shows the results are obtained in good agreement with literature data [11, 15],

These data are completed with the FTIR data, the best results being obtained for powder calcined at 1000°C [16]. The ion stretching vibration around 3568 cm^{-1} confirms the presence of a hydroxyl group. It should be noted that the hydroxyapatite sample (HAp1) calcinated at 800°C contains water of crystallization and hydration, the HAp2 sample subjected to calcination at 1000°C has a decreased water content, and the HAp3 sample calcinated at 1200°C contained a higher water content than HAp1 due to high reactivity of the powder [17].

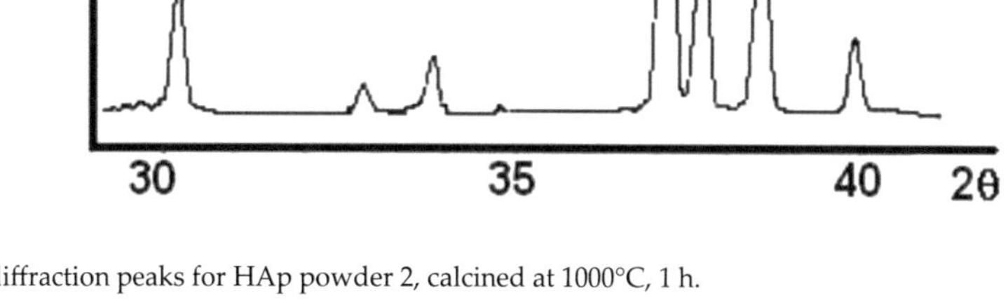

Figure 1. X-ray diffraction peaks for HAp powder 2, calcined at 1000°C, 1 h.

Likewise, the other stretching vibrations for carbonyl and phosphate groups were also observed as reported earlier [18, 19]. The broad absorption band from 1050 cm^{-1} is an evidence for tricalcium phosphate as reaction product. Literature studies indicate that tricalcium phosphate is present over 850°C. **Figure 2** showed the three FTIR curves performed on three types of hydroxyapatite, calcined at temperatures of 800, 1000, and1200°C, respectively. FTIR tests revealed the role of hydrogen and -OH group in the hydroxyapatite samples, and the only band that differentiate these three samples is the intense band from 1624.7 cm^{-1}.

Dynamic light scattering is a well-established, versatile, and noninvasive technique which can provide information about the size distribution of the particle populations in real time. DLS data provide information about the size distribution of the HAp crystals and their concentrations in aqueous solutions that is not available from other techniques [20]. It has also been extensively used to examine the sample homogeneity. The dynamic light scattering experiment shows that the particle size distribution is in the range of 50–70 nm, which is well supported by other techniques. Dynamic light scattering is used to monitor the size of the precipitating particles and to provide information about their concentration, including different associations in different orders, which generates large aggregates (**Figure 3**).

Structural characterization of hydroxyapatite powders calcinated at 1000°C was performed by scanning electron microscopy (SEM) and atomic force microscopy (AFM), too, for analyzing the morphology and particle size. In the SEM images some small crystallites of

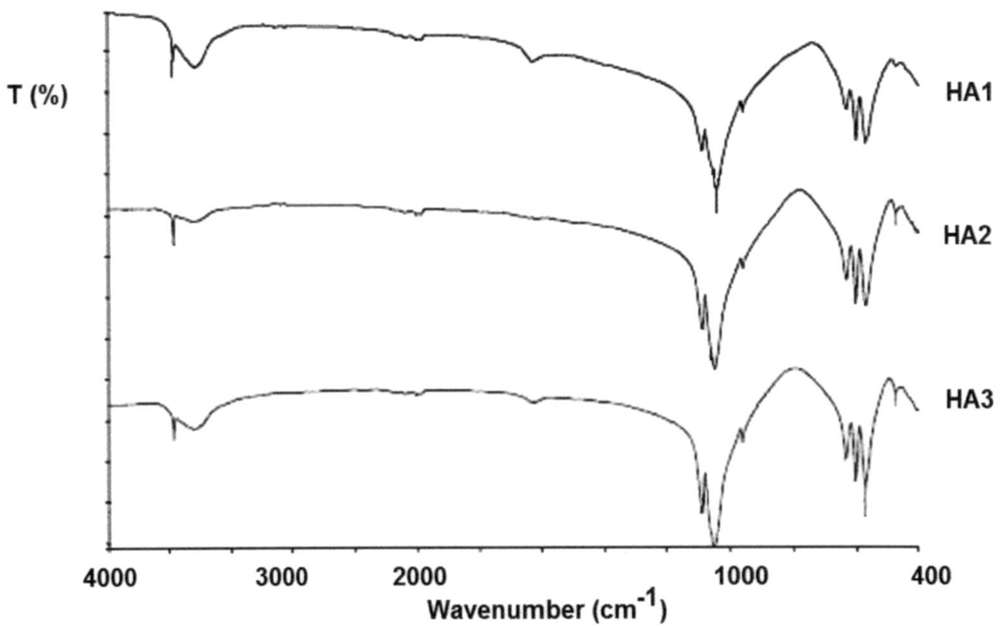

Figure 2. FTIR curves performed of hydroxyapatite, calcined at temperatures of 800, 1000, and 1200°C.

hydroxyapatite (<100 nm) and agglomerated particles are distinguished. Crystallites have uniform sizes with narrow particle size distribution.

The hydroxyapatite powder obtained by wet precipitation calcinated at 1000°C has been characterized by scanning electron microscopy. **Figure 4** presents scanning electron microscopy of hydroxyapatite sample calcined at 1000°C with a magnification of 2000×, and in **Figure 5**, it presents the same sample but at a magnification of 1000×. We can distinguish microporosity (<10 μm), allowing diffusion of ions and fluid from macroporosity (100–600μm) and can promote cell colonization.

To investigate with the atomic force microscope, hydroxyapatite solutions were freshly prepared before each experiment, by suspending an appropriate amount of each sample in ethanol. Compared with existing literature data [20], hydroxyapatite has a y-plane orientation. In our experiments, it can be observed that the grain is oriented occurring after z-plane.

By atomic force microscopy investigations, it was found that hydroxyapatite from bulk phase, at higher calcination temperatures, is agglomerated in nanoparticle phase and then in nanocrystalline spherule forms. Crystal size distribution depends on the critical size of nuclei in saturation conditions rather than the crystal growth if clusters of small particles are observed in AFM images.

In conclusion, hydrothermal synthesis method applied to HAp leads to the hydroxyapatite powders with a nanocrystalline of high and good stoichiometry, with a relatively narrow distribution of crystal size. Also, at high calcination temperature and long-term (4 h) of treatment, well-crystallized products were obtained, with desired parameters [21].

AFM technique has been used for evaluation the distribution of high spherules, visible in HAp powder [21]. The AFM method revealed a rugged HAp surface with crystallites ranging from 70 to 100 nm in accordance with other spectral methods. It has been found that at higher calcinations of temperatures HAp is deagglomerated from the bulk phase and agglomerates

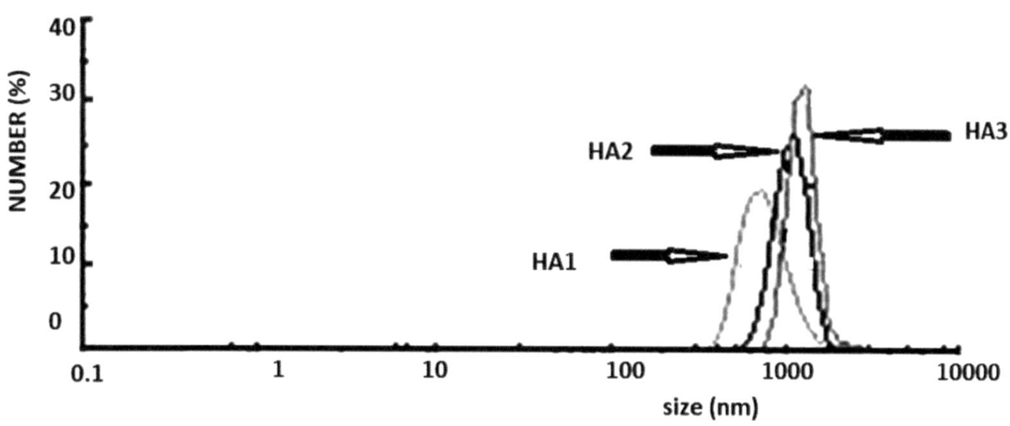

Figure 3. Particle size distribution of the three records by number.

Figure 4. Scanning electron microscopy of HAp powder calcined at 1000°C, 5000×.

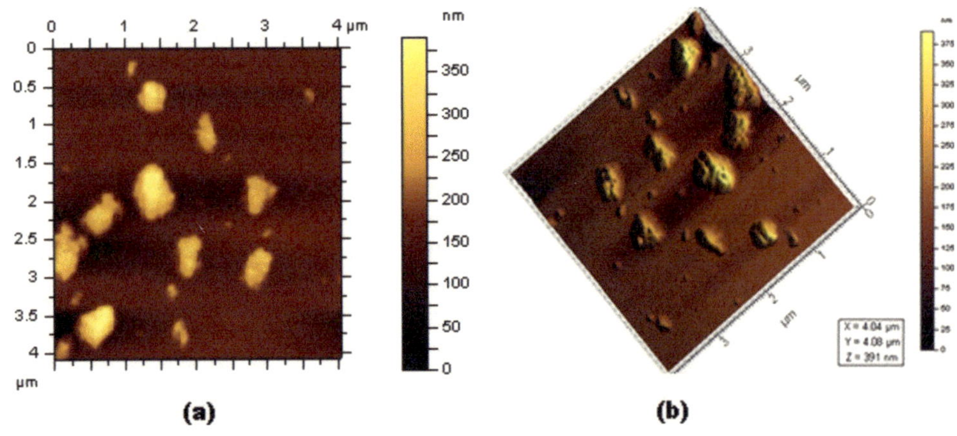

Figure 5. 3D representation of the grains of calcined hydroxyapatite at 1000°C.

in the nanoparticle phase that results in the formation of nanocrystalline and then spherules. The AFM 3D topography of hydroxyapaptite powder was recorded on an area of 5 × 5 μm^2, **Figure 5 (a,b)**.

The crystal size distribution depends on the critical dimensions of the nuclei under over-saturation conditions rather than on the crystal growth if particle agglomerations are observed.

2.2. Sol-gel method

The precursors used to prepare hydroxyapatite were calcium nitrate tetrahydrate $Ca(NO_3)_2$ $4H_2O$ and phosphorus pentoxide (P_2O_5) (Chimreactiv, Romania). To a solution of 0.5 mol/l, phosphorus pentoxide (P_2O_5) was dissolved in ethanol. $Ca(NO_3)_2.4H_2O$ was also dissolved in absolute ethanol to form a solution of 1.67 mol/l mixing both solutions were constantly stirred with a magnetic stirrer, after which the mixture was placed in an oven at an 80°C temperature for 24 h to complete the reaction [22, 23]. Dried gel was performed in an oven. The resulting product was a transparent gel. The samples were heat treated to transform the acid gel into a solid product with properties of biocompatibility and osteoconductivity

Before depositing onto 316L stainless steel, the hydroxyapatite film has been characterized by X-ray diffraction, and the results were relevant in terms of purity and presence of this compound. After analyzing, the diffraction phases identified in hydroxyapatite obtained by sol-gel were hydroxyapatite $(Ca)_{10}(PO_4)_6(OH)_2$ and tricalcium phosphate $Ca_3(PO_4)_2$. X-ray diffraction analysis reveals a high degree of crystallinity for HAp obtained by sol-gel. A broad reflection peak appears in the range of 31.8–32.51 2Θ, which represents the characteristic peak of apatitic phase (according to JCPDS card #9–432). Some characteristic peaks at, for instance, (211), (300), and (212) planes were shown for coatings annealed at higher temperatures 400–500°C. This suggests that the apatite coatings with structural evolution from amorphous to crystalline are able to produce depending on the temperatures.

3. HAp:316L composite materials

3.1. Composite HAp:316L obtained by cold pressing

Three technological versions of biocomposites obtained by powder metallurgy have been prepared by cold pressing, by varying the bulk concentrations [17, 18]. There have been five series of powder mixtures:

- Hydroxyapatite:316L stainless steel = 80%:20%.
- Hydroxyapatite:316L stainless steel = 50%:50%.
- Hydroxyapatite:316L stainless steel = 30%:70%.
- Hydroxyapatite:316L stainless steel = 20%:20%.

Quantitative measurement of the diameters of 316L powder particles was performed from SEM images. **Figure 6** shows SEM micrography of 316L stainless steel powder, and the size of powder varies between 142 and 195 μm. Powders are spherical slightly dendritic. There is a considerable disparity in both powder sizes which could lead to segregation during compaction, thus providing a compact product heterogeneity. One of the most important problems in

Figure 6. Scanning electron microscopy (500×).

the compaction process and its efficiency is given by applied powder shape. Hydroxyapatite powder is spherical, while the steel powder is dendritic in good agreement with literature's report [24, 25].

3.1.1. HAp:316L composite (80% HAp:20% 316L)

80% hydroxyapatite powder and 20% 316L stainless steel powder were mixed, were homogenized, and after which they were cold uniaxially pressed. The samples were sintered at 850 and 1000°C, held for 60 and 120 min. Sintering process took place in a protective atmosphere of H_2 with a cooling and heating at rate of 10°C/min (**Figure 7**).

Figure 7. Visual aspect of sample 2, 850°C, 120 min.

Macroscopic characterization of all four samples indicates a high porosity, exfoliation, and friability. Without lubricant for pressing, to avoid contamination with toxic material, a poor compaction of materials and a high porosity of the tablet are observed.

3.1.2. HAp:316L composite (50% HAp:50% 316L)

The material obtained by mixing powder of 50% 316L and 50% HAp was cut along the atomic planes. Material destruction was caused by different granulations of the two sorts of particles (too small for hydroxyapatite fraction <45 μm), the manual mixing and different densities (ϱ_{HAp} = 1.5 g/cm^3; ϱ_{316L} = 3.0 g/cm^3), of the two materials.

3.1.3. HAp:316L composite (70% 316L:30%HAp)

The preparation procedure of these composites was similar to the previous work on the same parameters. Fractions chosen for powder mixtures 70:30 were >160 μm for stainless steel 316L and >45 μm for hydroxyapatite. Three composite samples with cylinder shape have been obtained.

The SEM image of the obtained composite *HAp:316L composite (70% 316L:30%HAp)*, **Figure 8**. A relatively uniform distribution of stainless steel powder can be explained by mixing two types of powders performed manually. The metal particles looks as white formations, round, slightly dendritic, meanwhile the hydroxyapatite particles become dark coloured.

EDAX analysis of the composite with 70% 316L and 30% HAp is shown in **Figure 9**; the presence of phosphorus and calcium from hydroxyapatite and the presence of elements such as Ni, Cu, Cr, and Fe as specific metal elements of 316 stainless steel could be observed.

Figure 8. Composite image 70% 316L and 30% HAp.

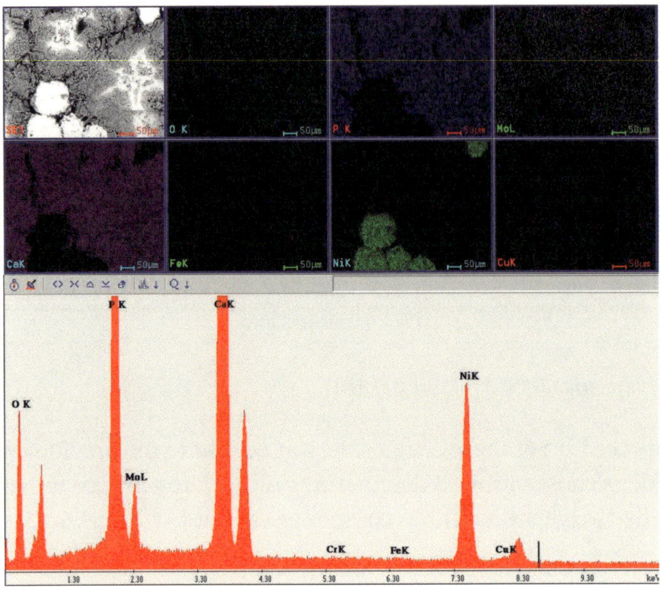

Figure 9. EDAX analysis for 70% 316L and 30% HAp composite.

3.1.4. HAp:316L composite (80%316L:20% HAp)

By a similar procedure, 80% 316L stainless steel powder and 20% of hydroxyapatite have been mixed. Two cylindrical composite with the dimensions, ϕ =10 mm and h = 13 mm, have been obtained. Composites obtained by mixing proportions of 80% 316L and 20% HAp powder are shown in **Figure 10**.

Figure 10 presents SEM microstructure of the 80% 316L and 20% HAp composite (images taken at 200× magnification). From microscopic analysis of the composite can be seen that the steel powder is the party round, slightly dendritic, white color. For these powder mixtures 80/20 were chosen fractions larger than 160 μm for 316L stainless steel powder and more than 45 μm for hydroxyapatite.

Figure 10. Composite image with 80% 316L and 20% HAp.

http://dx.doi.org/10.5772/intechopen.71490

EDAX analysis of 80% 316 and 20% HAp composite shown in **Figure 11** gives us the qualitative analysis of diffusion elements present in the composite. The diagram reveals the presence of phosphorus and calcium from hydroxyapatite and the presence of specific metal elements such as Ni, Cu, Cr, Fe, and Mo elements present in stainless steel 316L.

From all elements present in the chemical composition of 316L stainless steel, there is a moderate but uniform distribution of Fe and a reduced distribution of Ni. Distribution of Cu atoms is relatively uniform but stronger in specific areas of steel metal pellets.

3.1.5. Shore hardness determination

Shore D hardness (HSD) is a dynamic feature and represents an amount proportional to the ratio of the height at which jumps back and the height from which the material was dropped through a vertical guide. To determine hardness, composites are made by powder mixtures of 70% 316L:30%HAp and 80% 316L:20% HAp.

With the help of conversion tables, the Shore hardness values measured for both types of composites have been converted to HRC and HV values, as can be seen in **Table 1**.

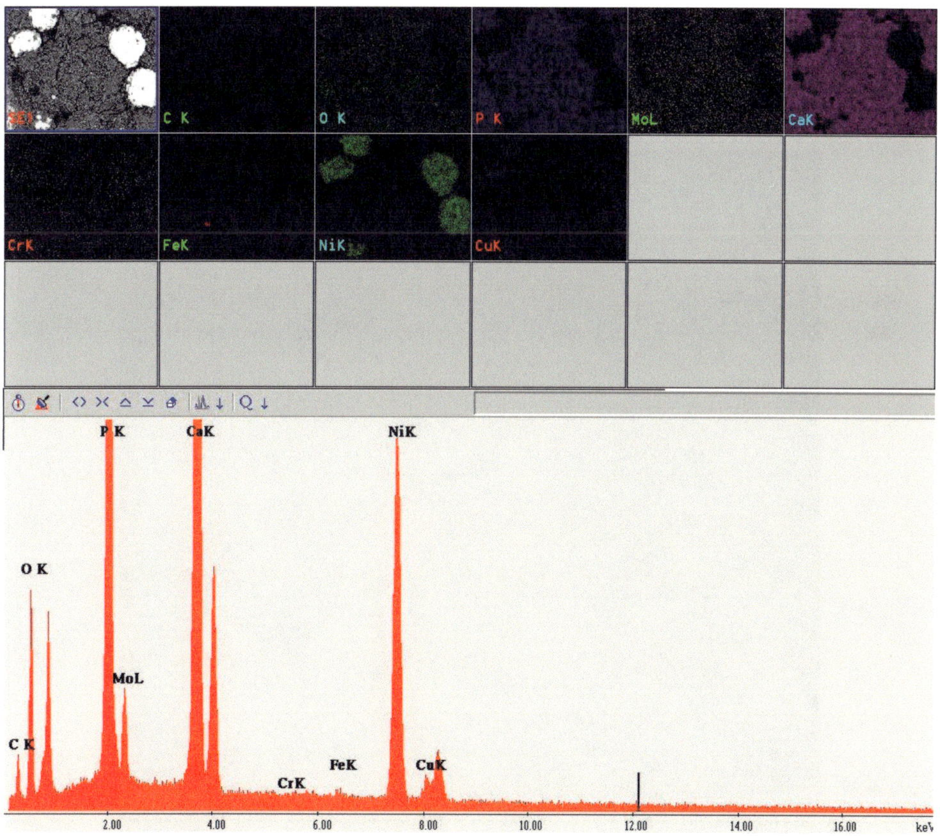

Figure 11. EDAX analysis for 80% 316L and 20% HAp composite.

No.	Composite	HSD	HRC	HV
1	70% 316L:30% HAp	71	53	560
2	80% 316L:20% HAp	77	57	643

Table 1. Composite hardness ratios with 30 and 20% HAp.

3.2. Hydroxyapatite coating of 316L stainless steel

A synthetic hydroxyapatite film obtained by sol-gel has been deposited on 316L stainless steel metal substrate, and the samples were subjected to X-ray diffraction analysis and structurally characterized by scanning electron microscopy. With this analysis the phases present in hydroxyapatite layer, morphology, and structure of hydroxyapatite were identified [24–27].

Structural quality of layers deposited by sol-gel was investigated by X-ray diffraction (XRD) analysis and scanning electron microscope (SEM) to highlight the crystallinity of deposited layers and the phases present in the hydroxyapatite film [4, 28, 29].

Figures 12 and **13(a, b)** show SEM images of hydroxyapatite layer deposited on 316L stainless steel. SEM photographies reveal the existence of a layer with a thickness about 350–500 nm and with an aspect of continuous film crack with a roughness of approximately zero [30, 31].

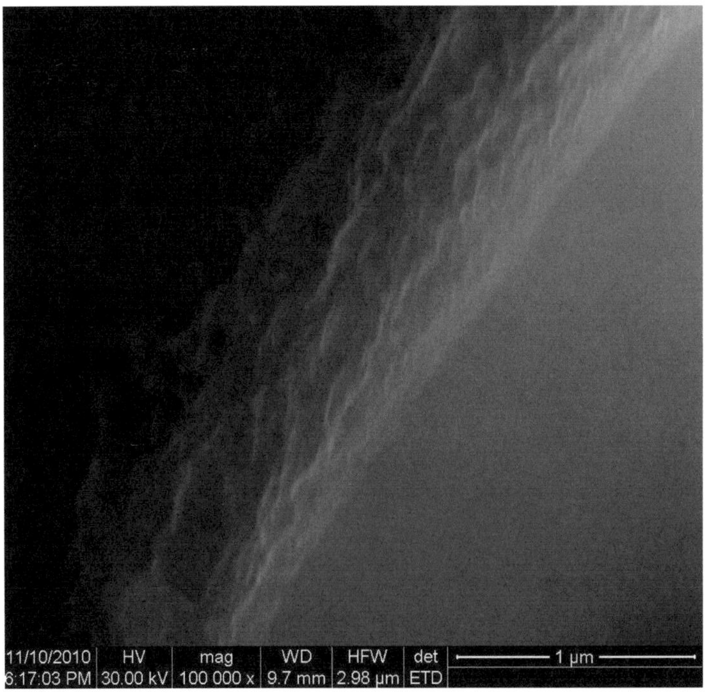

Figure 12. Film thickness of HAp deposited on 316L stainless (400,000×).

http://dx.doi.org/10.5772/intechopen.71490

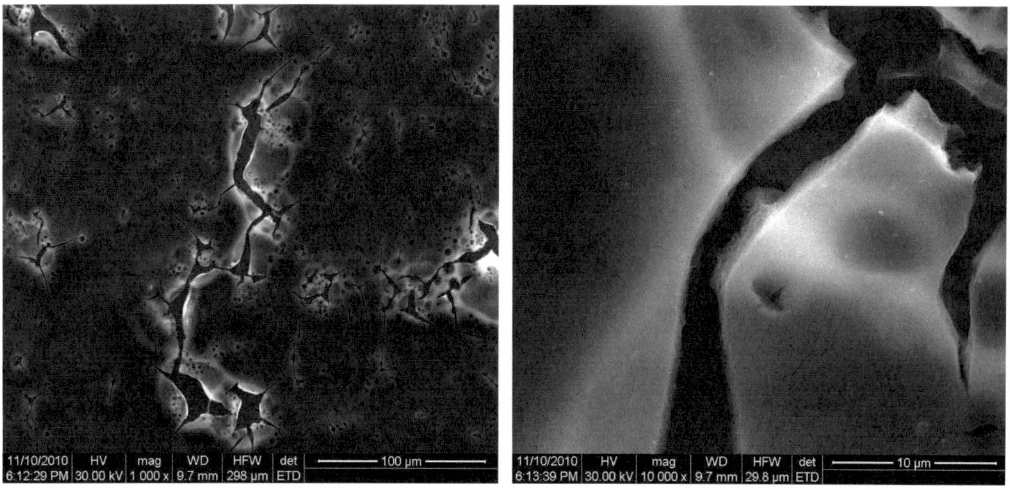

(a) (b)

Figure 13. SEM film of HAp (500×) (a) and SEM film of HAp (500×) (b).

Appearance of the two images reveals the presence of cracks due to burning contraction between steel and hydroxyapatite layer. SEM image of the first layer of HAp is shown in **Figure 14(a)** with 2500× magnification.

Appearance of the second layer of hydroxyapatite on 316L stainless steel support is shown in **Figure 14(b)** at 50,000× magnification. Phase analysis was performed on hydroxyapatite layer deposited on the metal by using X-ray diffraction (XRD) (**Figure 15**)

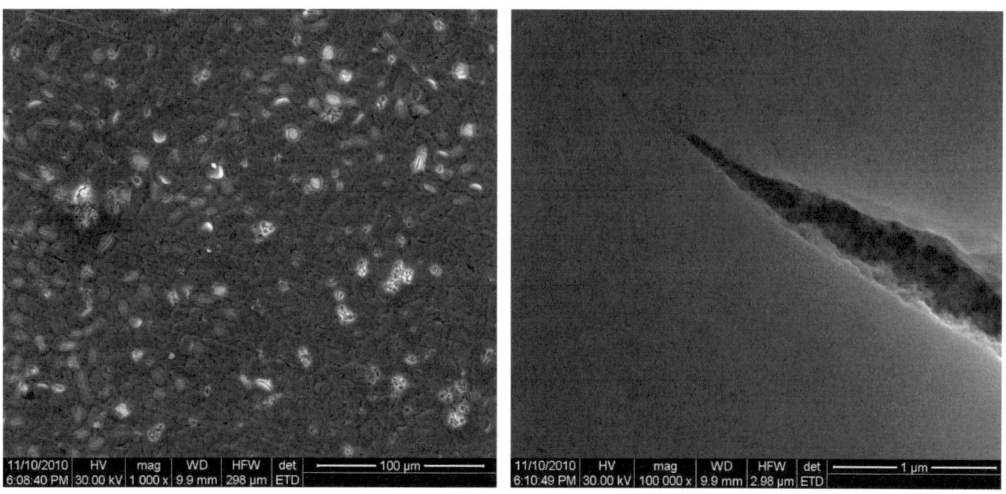

(a) (b)

Figure 14. SEM film of HAp on oxidized steel (2500×) (left) and SEM image of the second film (50,000×) (right).

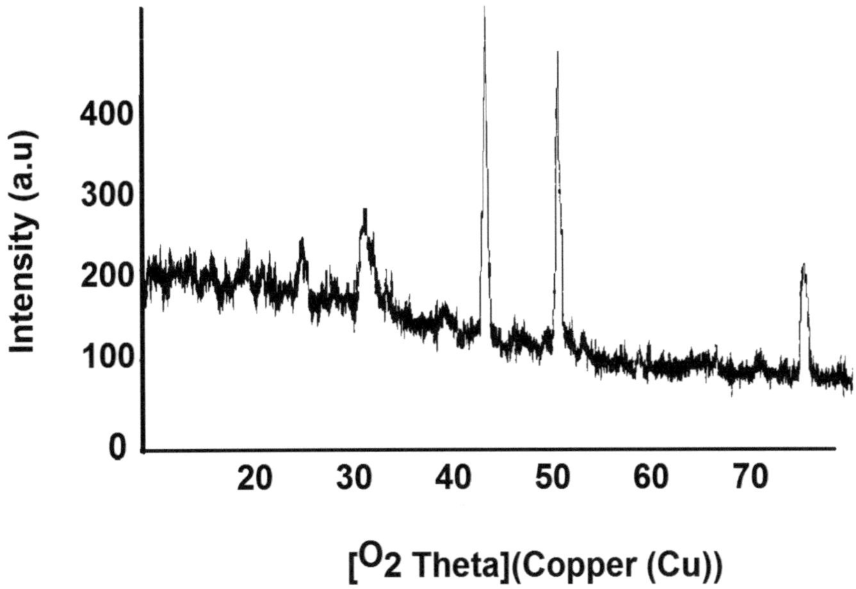

Figure 15. XRD diagram for hydroxyapatite layer deposited on the metal.

The only phase identified in the film deposited on 316L stainless steel substrate was hydroxyapatite. This method allows a good control of composition and crystallization of hydroxyapatite films at low temperatures [32, 33].

4. Behavior of the biomaterials in simulated physiological liquids

4.1. Determination of Fe, Mn, Ni, and Cr in saline and plasma by atomic absorption spectrometry

Stainless steel in medical applications is type 316 (AISI). In the 1950s, the amount of carbon in the type 316 was reduced from 0.08 to 0.03% by weight for a higher corrosion resistance in chlorides. This new alloy was known as the 316L. Even so, the use of austenitic stainless steels is limited due to the release of corrosion products in the form of Ni^{2+}, Cr^{3+}, and Cr^{6+}, which produce local effects in the body and destroy the implant [34].

For hydroxyapatite coating achieved by sol-gel method, two samples of 316L stainless steel sheet metal with 0.35 mm thickness with 15×15×0.35 mm dimensions were chosen. These samples were polished with metallographic papers 400 or electrochemically attacked with 5% perchloric acid in ethanol and then were washed and dried.

The relevant metal concentration that may be released from composite materials in plasma/serum, under controlled conditions in the ppbs, has not been reported too much in the literature; for the systems created and investigated in this chapter, the metal levels in plasma/serum have been analyzed by atomic absorption spectrometry and graphite furnace technique [35, 36].

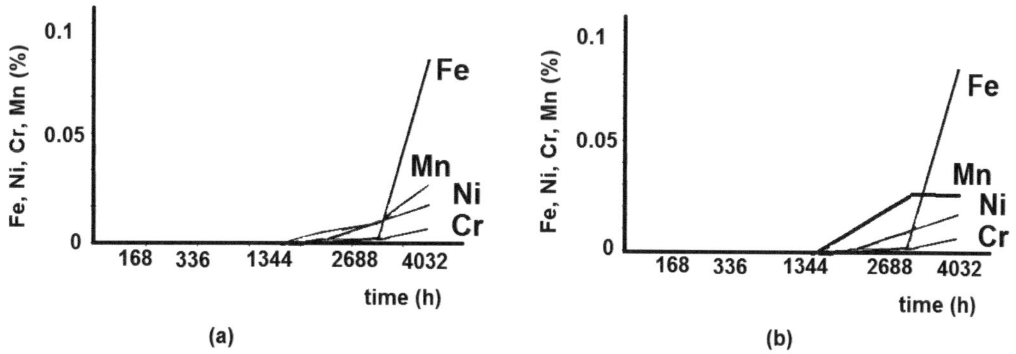

Figure 16. Changes in concentrations of Fe, Ni, Cr, and Mn in different samples.

Changes in concentrations of heavy metals like Fe, Ni, Cr, and Mn for the representative samples are plotted in **Figure 16** for two representative samples. These samples were chosen due to their higher surface exposure. Could be observed that after 6 months, the concentration of the metals (Fe, Mn, Cr, and Ni) (ppb) that may be released in simulated physiological fluids (SBF) from the used composite materials is practically zero.

The obtained results indicated that deposition of hydroxyapatite film on 316L metal plate was designed to prevent the release of metal ions in artificial plasma and in saline solution [37, 38]. **Figure 16a** presents changes in concentrations of Fe, Ni, Cr, and Mn in artificial plasma solution and in **Figure 16b** changes in concentrations of metals in saline solution.

5. Conclusions

In this chapter, the size and microscopic characterization was performed both for stainless steel powders and for hydroxyapatite and their composite. As compactness, the best compositions have been observed for 80% HAp: 20% 316L and 70% HAp: 30% 316L. The macroscopic analysis on composites with 80% HAp and 20% 316L obtained by cold pressing reveals friable samples with high porosity. It was found that:

- By increasing the amount of HAp in the composite, a lower hardness and a lower relative density for 316L/HAp composite, but with a significant increase in porosity, could be obtained.
- By increasing the proportion of HAp, the hardness or densification of the sintered composite is strongly affected.

The best mechanical results were obtained for composites of 70% 316L/30% HAp and 80% 316L/HAp 20%. Mechanical behavior studies of composites obtained were performed by analysis of Shore hardness, values of 71 HSD for composite with 70% 316L/30% HAp and 77HSD for composite with 80% 316/20% HAp. According to literature, for hydroxyapatite shore, the hardness values are between 81 and 88 HSD, and for human bones from 85 to 95 HSD. The results presented suggest that the composites obtained can be varied by choosing an appropriate chemical composition powder mixture.

The influence of grain size on the microstructure of the composite is very important because large granulations of powder will create large intergranular pores, which are affecting the density and mechanical properties of composites. HAp powder density proportional affects the product density (HAp/316L), so that a high density of particles will lead to a better packing density of compacted product.

In this chapter, compositional and structural investigations for composites have been established, and some correlations between the parameters of the process of preparation, chemical composition, structure, and properties of composites have been obtained, too. A comparison between composites and sol-gel coating is discussed, too.

Acknowledgements

For this chapter the authors received a financial support from UEFISCDI-MEN through the projects PNII 185/2014, PNIII 120 BG/2016, and PN 16.31.02.04.04.

Author details

Aurora Anca Poinescu[1] and Rodica-Mariana Ion[1,2*]

*Address all correspondence to: rodica_ion2000@yahoo.co.uk

1 Materials Engineering Department, Valahia University, Targoviste, Romania

2 ICECHIM, Nanomedicine Research Group, Bucharest, Romania

References

[1] Park JB, Bronzino JD. Metallic biomaterials. In: Taylor Print, editor. Park JB. Biomaterials: Principles and Applications. Boca Raton, London, New York, Washington, D.C.: CRC Press; 2013

[2] BomBac D, Brojan M, FajFar P, Kosel F, Turk R. Review of materials in medical applications. RMZ-Materials and Geoenvironment. 2007;**54**(4):471-499

[3] Marchant RE, Wang I. Physical and chemical aspects of biomaterials used in humans. In: Implantation Biology: The Host Response and Biomedical Devices. 1994. p. 13-38

[4] Fan X, Chen J, Zou J, Wan Q, Zhou Z, Ruan J. Bone-like apatite formation on HA/316L stainless steel composite surface in simulated body. Transactions of Nonferrous Metals Society of China. 2009;**19**:347-352

[5] Ruan JM, Zou JP, Zhou ZC. Hydroxyapatite-316L stainless steel fibre composite biomaterials fabricated by hot pressing. Powder Metallurgy. 2006;**49**(1):62-65

[6] Poinescu AA, Ion RM, Van Staden RI. Different microscopic characterization techniques on hydroxyapatite powder. Journal of Optoelectronics and Advanced Materials. 2011;**13**(2-4):416-421

[7] Monmaturapoj N. Nano-size hydroxyapatite powders preparation by wet-chemical precipitation route. Journal of Metals Materials and Minerals. 2008;**18**(1):15-20

[8] Sung YM, Shin YK, Ryu JJ. Preparation of hydroxyapatite/zirconia bioceramic nanocomposites for orthopaedic and dental prosthesis applications. Nanotechnology. 2007;**18**: 65602-65607

[9] Poinescu AA, Ion RM, Trandafir I, Bacalum E, Radovici C. Obtaining and Characterization of a Calcium HA. In: Proc.TEHNOMUS XV "New Technologies and products in machine manufacturing technologies; 2009. p. 301-306

[10] Poinescu AA, Ion RM, Van Staden RI. Proceedings of SPIE-the International Society for Optical Engineering, investigations on hydroxyapatite powder obtained by wet precipitation. In: SPIE - the International Society for Optical Engineering. Advanced topics in optoelectronics, microelectronics, and nanotechnologies V; August 2010; Constanta, Romania. 2010. p. 78210B1-78210B6

[11] Slosarczyk A, Paszkiewicz Z, Paluszkiewicz C. FTIR and XRD evaluation of carbonated hydroxyapatite powders synthesized by wet methods. Journal of Molecular Structurc. 2005;**744-747**:657-661

[12] Sahin E. Synthesis and characterization of hydroxyapatite-alumina-zirconia biocomposites [thesis], Izmir; 2006. p. 75

[13] Ciobanu CS, Andronescu E, Stoicu A, Florea O, Le Coustumer P, Galaup S, Djouadi A. Influence of annealing treatment of nano-hydroxyapatite bioceramics on the vibrational properties. Digest Journal of Nanomaterials and Biostructures. 2011;**6**(2):609-624

[14] Donadel K, Laranjeira CMM, Gonçalves VL, Favere VT, De Lima JC, Prates LHM. Hydroxyapatite produced by wet-chemical methods. Journal of the American Ceramic Society. 2005;**88**(8):3374-3374

[15] Cengiz B, Gokce Y, Yildiz N, Aktas Z, Calimli A. Synthesis and characterization of hydroxyapatite nanoparticles. Colloids and Surfaces A: Physicochemical and Engineering Aspects. 2008;**322**(1-3):29-33

[16] Rintoul L, Wentrup-Byrne E, Suzuki S, Grøndahl L. FT-IR spectroscopy of fluoro-substituted hydroxyapatite: Strengths and limitations. Journal of Materials Science. Materials in Medicine. 2007;**18**(9):1701-1709

[17] Rehman I, Bonfield W. Characterization of hydroxyapatite and carbonated apatite by photo acoustic FTIR spectroscopy. Journal of Materials Science. Materials in Medicine. 1997;**8**(1):1-4

[18] Boanini E, Bigi A. Biomimetic synthesis of carbonated hydroxyapatite thin films. Thin Solid Films. 2006;**497**:53-57

[19] Bogza ES. Studii cinetice şi de echilibru ale unor procese de reţinere pe materiale apatitice [thesis]. Cluj Napoca; 2010

[20] Poinescu AA, Ion RM. Particles dimensional analysis and microscopic characterization of hydroxyapatite powder. In: TEHNOMUS—New Technologies and Products in Machine Manufacturing Technologies; 2011; Suceava. Suceava: Universitatea "Stefan Cel Mare", Suceava; 2011. p. 283-288

[21] Chai C, Ben-Nissan B. Thermal stability of synthetic hydroxyapatites. International Ceramic Monographs. 1994;**1**(1):79-85

[22] Chai CS, Ben-Nissan B. Bioactive nanocrystalline sol-gel hydroxyapatite, coatings. Journal of Materials Science. Materials in Medicine. 1999;**10**:465-469

[23] Anderson D, Hastings GW, Morrey S, Rich C, Hydroxyapatite ceramic coatings. In: Heimke G, editor. Bioceramics, vol 2. Proceedings of the 2th International Symposium on Ceramics in Medicine; Heidelberg: 1989. p. 251

[24] Berndt CC, Haddad GN, Gross KA. Thermal spraying for bioceramics application. In: Heimke G, editor. Proceedings of the 2th International Symposium on Ceramics in Medicine; Heidelberg: 1989. p. 201

[25] Ben-Nissan B, Milev A, Vago R. Morphology of sol-gel derived nano-coated coralline hydroxyapatite. Biomaterials. 2004;**25**:4971-4975

[26] Ion RM. Nanocrystalline Materials. Bucharest: FMR; 2003. 189 p

[27] Poinescu AA, Radulescu C, Vasile BS, Ionita I. Research regarding sol-gel hydroxyapatite coating on 316L stainless steel. Revista de Chimie. 2014;**65**(10):1245-1248

[28] Ballarre O, Liu Y, Mendoza E, Schell H, Diaz F, Orellano JC, Fratzl P, Ceré S. Enhancing low cost stainless-steel implants: Bioactive silica-based sol-gel coatings with wollastonite particles. Journal of Nano and Biomaterials. 2012;**4**:33-53

[29] Bermudez R, Espinoza Beltran FJ, Espitia Cabrera E, Contreras Garcia ME. Characterization of HAp-ZrO2 base bilayer on 316L stainless steel substrates for Orthopaedic prosthesis applications. Advanced in Technology of Materials and Materials Processing Journal. 2007;**9**(2):141-148

[30] Cao W, Hench LL. Bioactive materials. Ceramics International. 1996;**22**:493-507

[31] Miao X. Observation of microcracks formed in HA-316L composites. Materials Letters. 2003;**57**:1848-1185

[32] Nayar S, Hydroxyapatite PA. Coating on stainless steel pre-coated with bovine serum albumin at ambient conditions. Colloids and Surfaces B: Biointerfaces. 2006;**48**:183-187

[33] Seo DS, Kim YG, Hwang KH, lee JK. Preparation of hydroxyapatite powder derived from tuna bone and its sintering property. Journal of the Korean Ceramic Society. 2008;**45**(10):594-600

[34] Balamurugan A, Balossier G, Kannan S, Michel J, Faure J, Rajeswari S. Electrochemical and structural characterisation of zirconia reinforced hydroxyapatite bio ceramic sol–gel coatings on surgical grade 316L SS for biomedical applications. Ceramics International. 2007;**33**:605-614

[35] Wataha JC, Nelson SK, Lockwood PE. Elemental release from dental casting alloys into biological media with and without protein. Dental Materials. 2011;**17**:409-415

[36] Williams DF. The Williams Dictionary of Biomaterials. 1st ed. Liverpool: Liverpool University Press; 1999

[37] Aizawa T, Kaneko T, Yajima H, Yamada S, Sato Y, Kanda Y, Kanda S, Noda M, Kadowaki T, Nagai M, Yamauchi K, Komatsu M, Hashizume K. Rapid oscillation of insulin release by the rat pancreatic islets under stringent Ca^{2+}-free conditions. The Journal of Endocrinology. 2000;**166**:545-551

[38] Poinescu AA, Ion RM, Stan R, Rizescu CZ. Determining sensitivity to intergranular corrosion of austenitic stainless steel 316L. The Scientific Bulletin of Valahia University-Materials and Mechanics. 2010;**5**(8):91-94

Chapter 9

Gene Delivery by Hydroxyapatite and Calcium Phosphate Nanoparticles: A Review of Novel and Recent Applications

Feray Bakan

Additional information is available at the end of the chapter

http://dx.doi.org/10.5772/intechopen.71062

Abstract

Gene therapy is a targeted therapy which can be used in the treatment of various acquired and inherited diseases. Inhabitation of a gene function, restoring or improving a gene, or gaining a new function can be achieved by gene therapy strategies. The most crucial step in this therapy is delivering the therapeutic material to the target. Nanosized calcium phosphates (CaPs) have been considered as promising carriers due to their excellent biocompatibility. In this chapter, the delivery of DNA, siRNA, and miRNA by using CaP nanocarriers were compiled in detail and the main parameters which can affect the carrier properties and thus the gene transfer efficiency were also discussed.

Keywords: calcium phosphate nanoparticles, non-viral vector, hydroxyapatite, gene therapy, gene silencing

1. Introduction

Recently, targeted therapies are becoming more attractive since they do not harm healthy cells and have high selectivity. Gene therapy is the modification of patients' cells genetically for curing or making the health state better. With gene therapy strategies, one gene function can be inhibited, restored or improved or a new function can be gained. For an improvement in functional disorders induced by the mutation or to inhibit the expression of an inadequate gene, a therapeutic gene material, and its related regulatory components are delivered to the nucleus to treat the diseases. In most cases, owing to their negative charge, naked oligonucleotides cannot penetrate the cell; therefore, an appropriate carrying agent is needed. The most crucial issue in gene therapy is delivering the therapeutic gene, which first must overcome the extracellular barriers and, subsequently, the cellular barriers. The carrier used for loading them is called "a vector," while the vectors recently been used for gene therapy are roughly sorted as viral vectors and non-viral vectors. However, viral vectors' side effects, potential cytotoxicity of the carriers, and poor transfection efficiency of non-viral vectors have currently limited the accomplishment of gene therapy. As the use of nanoparticles has been a remarkable methodology in the solution of a variety of problems, they can be employed as non-viral delivery vehicles for oligonucleotides in molecular biology and medicine [1]. With having excellent biocompatibility and high chemical affinity toward DNA and RNA, calcium phosphate nanoparticles can participate in the delivery systems which can limit most of the problems in delivering the therapeutic molecules to the nucleus of target cells. Further, calcium phosphates can overcome the extracellular barriers and then dissolve in the acidic pH within endosomes and lysosomes which lead the release of nucleic acids in the targeted region of the cell [2].

This review focuses on the potential employment of hydroxyapatite and other calcium phosphate nanoparticles as non-viral vectors in gene therapy and gene silencing as well as emphasizing the recent studies to expose the benefits for using such vectors.

2. What is gene therapy?

The genes on the chromosomes are the smallest genetic units that are effective in producing vital proteins for the cells. Any mutation that occurs in the genome can cause the disease because it changes the protein functions. Gene therapy is basically described as the transfer of genetic material with therapeutic effect to specific targeted cells or tissues with minimal toxicity to provide clinical benefits in the treatment of genetic and infectious diseases. Inhabitation of a gene function, restoring or improving a gene, or gaining a new function can be achieved by gene transfer strategies.

Gene therapy has been quite important in treating the various acquired and inherited diseases (e.g., AIDS, cancer, and other genetic disorders). As a theory, if the sequence of a mutant gene which is defined as an oncogene is recognized, gene therapy can be used in the treatment of any cancer disease. Moreover, its execution in the cure of genetic disorders such as muscular dystrophy, cystic fibrosis and severe combined immunodeficiency (SCID) has attracted attention due to three main reasons: (a) being well-characterized, (b) knowing the mutations responsible for the disease, and (c) existing no other satisfactorily efficient treatment [3].

Independent of the type of applied gene therapy method, the therapy first identifies the mutant gene that causes the disease. Following that, the identical healthy gene (e.g., therapeutic gene or transgene) is cloned. The therapeutic gene is designed according to the augmenting or suppressing or repairing needs. Since the penetration of "naked" oligonucleotides into cells is not probable, an appropriate carrier must be used in loading the produced therapeutic gene [4, 5]. This carrier is called vector and it is responsible for transporting the therapeutic gene to the patient's target cell successfully. In this sense, a vector needs to come through the extracellular barriers and, then, the cellular barriers [6, 7]. Consecutive stages in gene therapy are given in **Figure 1**. The biggest challenge in a successful gene therapy is the issue of

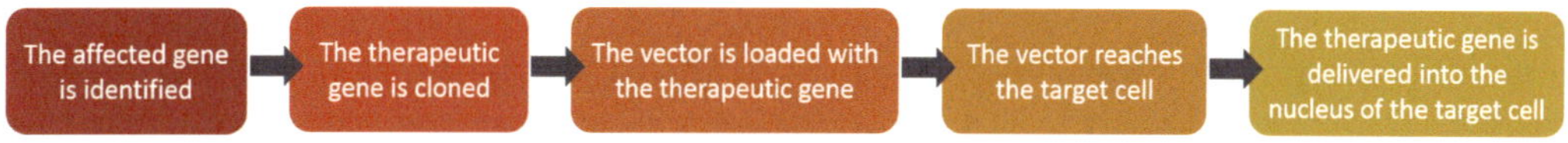

Figure 1. Schematic illustration of consecutive stages in gene therapy.

delivery. The optimal vector and the delivery system depend on the target cells and its characteristics, duration of expression, and the size of the genetic material to be incorporated in the vector [8]. Therefore, an optimal delivery vector needs to be safe, particular, and efficient.

There are two main sorts of vectors used in the gene therapy: viral vectors and non-viral (or synthetic) vectors. Viral vectors are viruses that have been genetically mutated for transporting normal human DNA. Viruses have developed a way of encapsulating and transferring their genes to human cells in a pathogenic manner. This capability can be used in the manipulation of the viral genome to replace disease-causing genes with the therapeutic ones [9]. Among all vectors, viral vectors exhibit higher efficiencies for transfecting host cells. However, the immunogenicity and the cytotoxicity are the two major drawbacks of using these carriers [10]. The first failure in the clinical trial of the related gene therapy was identified with the inflammatory response to the viral vector (Adenovirus). An additional and very important reason for the concerns over the employment of the viral vectors is the formation of insertional mutagenesis in which an exogenous DNA sequence integrates into the genome of a host organism. This can trigger the proliferation of oncogenes which leads to the malignant transformation of cells [8]. Therefore, non-viral vectors have the safety advantage over the viral vectors due to the demonstrated reduced pathogenicity. However, due to low delivery efficiency, thus, poor expression of their transgenes, the applications of non-viral gene transfer were previously disregarded [8]. Since non-viral vectors have less immunotoxicity, they have been applied in clinical trials from 2004 to 2013, whereas the use of viral vectors in the same treatments decreased significantly, then. Applications of a variety of non-viral vectors entering clinical trials, increased because of the advances in their performance, selectivity, gene expression period, and safety characteristics [8]. For therapeutic purposes; DNA, mRNA and short double-stranded RNA, including small interfering RNA (siRNA) and microRNA (miRNA) mimics, can be delivered by using various non-viral vectors; however, there might be some drawbacks in the use of non-viral vectors in clinical trials. **Figure 2** presents the barriers to achieve a successful in vivo delivery of nucleic acids using non-viral vectors [11]. The degradation caused by serum endonucleases, the immune detection, the renal clearance from the blood, and the emerging nonspecific interactions are some of these handicaps and they need to be considered and avoided for a successful delivery. Furthermore, the carrier needs to extravasate from the bloodstream in order to reach the tissue targeted. Within the target tissue, it should mediate the cell entry and endosomal escape, as siRNA and miRNA mimics must be loaded into the RNA-induced silencing complex (RISC), whereas mRNA must bind to the translational machinery. DNA should be further transported to the nucleus to implement its activity [11].

Owing to their easy preparation and surface-functionalization, inorganic nanoparticles are promising non-viral vectors for gene therapy applications. If the nanoparticles are small enough (<100 nm), different nanoparticle materials such as calcium phosphates, polymers,

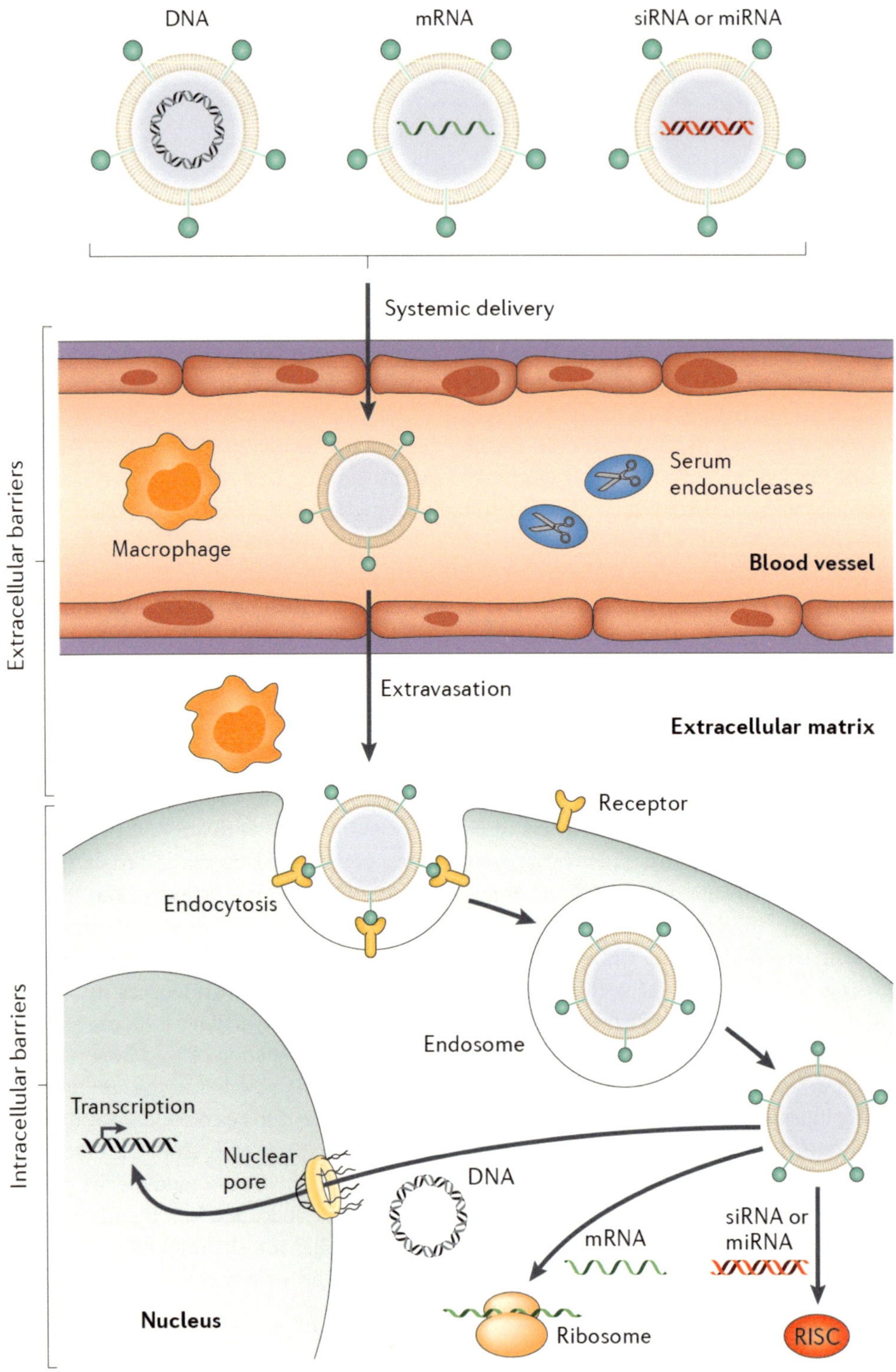

Figure 2. The barriers to success in vivo delivery of nucleic acids using non-viral vectors. Reprinted by permission from Macmillan Publishers Ltd.: [Nature Reviews Genetics] (13), copyright (2014).

liposomes, silica, magnetite, carbon, clay, and gold can be employed as non-viral vectors [8]. Compared to rest of all nanoparticles, calcium phosphate (CaP) has several advantages because of its easy preparation [12], high chemical affinity toward DNA and RNA [13, 14], high biocompatibility, and good biodegradability properties in biological systems [12]. Further, CaP particles have the ability to enter the cell and breakdown in the cell, also addressing them as an appealing candidate among all other gene delivery vectors [1, 2].

3. Calcium phosphates

CaP is one of the major mineral constituents of all mammalian calcified tissues and it is biocompatible, biodegradable, non-toxic, and non-immunogenic [14], therefore, CaP has widespread use in nanomedicine, including tissue engineering, imaging, and drug/gene delivery [14]. Furthermore, it can degrade into ions inside the early lysosomes of the cells, which makes it a convenient and safe intracellular delivery agent for the therapeutic small molecules and genes [14, 15]. Therefore, CaP nanoparticles have been widely used as non-viral vectors since the 1970s [16]. Negatively charged nucleic acids can bind to CaP nanoparticles with high affinity via calcium ion chelation and are consequently protected from degradation by endonucleases [14].

Hydroxyapatite (HA), with the formula $Ca_{10}(PO_4)_6(OH)_2$, has a Ca/P molar ratio of 1.67 and it is the most stable and least soluble phase of all calcium orthophosphates. Pure HA never forms in biological systems. Nevertheless, because of its excellent biocompatible, bioactive, and osteoconductive properties, HA is widely used in medical and dental applications, such as hard tissue repair, bone defect filling, bone regeneration, coatings of metal prostheses, and tumor surgery applications, etc. [17]. HA nanoparticles have recently been used in medical applications as carriers for growth factors, antibiotics, cancer drugs, antigens, and nucleic acids [18]. Moreover, HA columns are used for protein chromatography [19]. Several techniques, such as solid state, precipitation, sol-gel, hydrothermal, multiple emulsion, and biomimetic deposition have been used in the synthesis of HA nanoparticles [20]. However, some variability in the structure and morphology may occur depending on the synthesis method [21].

The non-stoichiometric apatites are of biological importance because they resemble mineral component of bone, enamel, dentin, and cementum. Bone mineral is essentially calcium-deficient hydroxyapatite (CDHA, $Ca_{10-x}(HPO_4)_x(PO_4)_{6-x}(OH)_{2-x}$) with a Ca/P ratio of approximately 1.5. While CDHA is chemically and compositionally similar to beta-tricalcium phosphate (β-TCP), it exhibits structural similarity to stoichiometric HA [14]. CDHA exhibits high solubility in water or body fluids while high crystalline stoichiometric HA has poor solubility. Over a broad range of concentration, pH, and temperature, non-stoichiometric apatites can be synthesized by various precipitation techniques [21].

3.1. The use of calcium phosphate nanoparticles for gene therapy

Although several studies have focused on the direct injection of naked DNA into various tissues such as skeletal muscle, liver, thyroid, heart muscle, brain, and urological organs, significantly (respectively) low transfection efficiencies were gained [22]. The literature indicates that there are three general levels that non-viral CaP-DNA delivery systems operate: (a) DNA condensation and complexation, (b) endocytosis, and (c) nuclear targeting or entry [23]. Before introducing into the cell, negatively charged DNA is complexed with CaP nanoparticles. Because of its positive charge, CaP-DNA complex gets bound to the negatively charged cell membrane and is consequently received by the cells. For the uptake of macromolecules or nanoparticles, endocytosis as a frequent methodology can internalize non-viral gene delivery vectors [24]. Endocytosis can be defined as the penetration of the cell membrane which is followed by the incorporation into an intracellular vesicle [25]. After endocytosis, the vector-DNA complex is mostly kept in perinuclear endosome/lysosomes. In case, DNA does not escape from the endosome, endosomal degradation of DNA might happen during endocytosis. The acidic pH (5–5.5) of endosomes and lysosomes usually causes the decomposition of DNA captured inside the vectors [23]. For this reason, DNA transfection requires endosomal escape. The degradation of DNA by particular enzymes (nucleases) depends on an efficient release of DNA into the cytoplasm. There are some barriers needed to overcome for having a successful entry of DNA into the nucleus; hence, protection from the nucleases is required. In the next step, the DNA is introduced into the nucleus. Nuclear pore complexes (NPCs) make the transfer of molecules into the nucleus [10]. Even though NPCs are very permeable to small molecules, they restrict the movement of larger molecules across the nuclear envelope. For the purpose of eliminating this obstacle; first, macromolecules carrying a nuclear localization sequence (NLS) are recognized by importins, and next they are actively transported through the pore into the nucleus [10]. Furthermore, Ca^{2+} regulated transport involves intermediate-sized molecules (10–70 kDa) and it does not entail an NLS [23]. Under conditions in which Ca is present in the cytosol (such as from CaP), the central plug is located well below the cytoplasmic ring of the NPC, and the pore can allow the diffusion of even medium-sized molecules. Under such circumstances, DNA-Ca complexes can easily enter the nucleus through the NPC.

The biochemistry field has been using the standard calcium phosphate transfection method, since it was first introduced by Graham and van der Eb in 1973 [26]. In many studies, calcium and phosphate solutions are mixed in the presence of DNA and coprecipitated with DNA to form the transfection complex [27]. However, this technique allows for the formation of a heterogeneous size distribution of CaP-DNA complexes and induces a large deviation in the transfection efficiency. The control of the growth and thus the size of the CaP-DNA complex are the two important factors for developing DNA delivery systems [28]. The size of the complex can be regulated by adding organic or inorganic modifiers during the precipitation process. Kakizava et al. [28] developed monodisperse CaP nanoparticles surrounded by hydrophilic polymer layer through complexation with poly(ethylene glycol)-block-poly(aspartic acid) (PEG-PAA) via basic mixing of calcium/DNA and phosphate/PEG-PAA solutions. It was shown that the use of PEG and PAA is beneficial to obtain a narrow size distribution. The efficiency of the complex was determined by measuring the amount of DNA incorporated into the particles by using gel permeation chromatography and fluorescence measurements. Similarly, HA nanorods with a length of 100 nm and a diameter of 25 nm were synthesized by using a stabilizing block copolymer (PLGA–mPEG) and the DNA loading capacity was investigated [29]. In this study, first, HA nanorods were synthesized and then, the HA-DNA complex was obtained by quick mixing them together in the presence of $CaCl_2$.

Plasmid DNA and salmon sperm DNA were used to test the DNA loading capacity and a huge enhancement of DNA binding capacity was observed. The strategy of using block copolymer and post-binding of HA nanorods with DNA make them remarkably efficient and safe gene vector [29]. Zhu et al. [30] synthesized short HA nanorods with a length of 40–60 nm and then mixed them with EGFP-N1 pDNA to form the nanoparticle-DNA complex. The agarose gel electrophoresis showed that the HA nanorods potentially adsorb EGFP-N1 pDNA under acidic (pH = 2) and neutral (pH = 7) conditions. HA-EGFP-N1 pDNA complex was transfected to the SGC-7901 cells in vitro with the efficiency about 80%. Two weeks after tail vein injection into mice, no acute toxicity was revealed. The liver, the kidney, and the brain organ tissue specimen of the sacrificed mice were investigated via TEM after receiving the injection of pDNA complex. EGFP green fluorescence expression was observed, which suggested that the EGFP-N1-pDNA did effectively transport on the HA nanoparticles into these organ tissues and the pDNA did integrate into the cell genome and ultimately successfully expressed there.

In another study, CaP nanoparticles with varying calcium (Ca) to phosphate ion (P) ratios were synthesized and the effect of the Ca/P stoichiometry on controlling the particle size and the plasmid DNA binding efficiency was investigated [31]. The results indicated that the determination of the Ca/P ratio which optimizes both DNA binding and condensation can overcome many barriers associated with transporting the DNA into the nucleus and thus improve transfection efficiencies. The reported optimal size of the complex is about 25–50 nm which performed efficiently at both binding and condensing the plasmid DNA.

To obtain a better capturing of the DNA, substitute ions, such as Mg^{2+}, CO_2^3, K^+, and Al^{3+}, can be added in CaP crystal lattice [32]. The studies concerning the effect of the substitute ions on the precipitation and the crystallization of HA and other CaPs are present in literature. It was reported that the integration of Mg^{2+} into the CaP lattice limits the particle growth and this leads to the formation of smaller sizes of CaP/DNA or CaP/siRNA precipitates, which resulted in remarkably enhanced cellular uptake. Chawdory et al. [33] reported that an increase in the dose of Mg^{2+} dramatically reduced the particle diameters and remarkably higher cellular uptake of DNA and its subsequent expression (N10-fold) was obtained compared with classical calcium phosphate coprecipitation [33].

Another strategy for improving the transfection efficiency is producing a multi-shell CaP-DNA complex, in which calcium phosphate core was coated with DNA, followed by calcium phosphate coating, and finally a second DNA coating [5, 34]. In this way, DNA can be better protected from enzymatic degradation and intracellular attack and therefore, a considerable increase in transfection efficiency can be obtained compared to single-shell CaP-DNA complex. It should be considered that the reaction temperature, the way of mixing, reagent concentration, pH, precipitation time, and the time between precipitation and transfection are important for the efficiency of cell transfection [1, 5, 34].

To protect the therapeutic agent from the biological environment, CaP-DNA can be encapsulated within a synthetic polymer such as polyethylene glycol, poly(methacrylic acid) or polyethyleneimine. However, synthetic polymers such as polyethyleneimine (PEI) are cytotoxic and not suitable for the in vivo delivery of nucleic acids. The toxicity of PEI partly comes from its limited biodegradability [35]. Klesing et al. [36] functionalized commercial hydroxyapatite nanorods with polyethyleneimine (PEI) and an average particle size of 190 nm without any agglomerates was obtained. In order to track the adsorption capacity of the HA-PEI nanorods, varying amounts of DNA was added to the dispersion and the dispersion stability was monitored by dynamic light scattering. Varying amounts of EGFP encoding DNA were added to the HA–PEI nanorods for the transfection of HeLa and MG-63 cells. As the addition of DNA reduces the surface zeta potential of the cationic HA-PEI delivery system, the nanorods which have small amounts of DNA exhibit higher positive zeta potential and thus better cellular uptake by the negatively charged cell membrane.

Since the cell membrane is negatively charged, positively charged nanoparticles lead to the higher degree of internalization due to the ionic interactions occurring between positively charged particles and cell membranes [37, 38]. Furthermore, it was reported that the positively charged nanoparticles might escape from lysosomes if they are internalized and show perinuclear localization, while the negatively and neutrally charged nanoparticles colocalized with lysosomes [39, 40]. The surface modification can strongly alter some characteristics of nanoparticles (e.g., hydrophilicity, hydrophobicity, and surface charge). Therefore, conjugating the nanoparticles with special functional groups can enhance the cellular uptake and thus the transfection efficiency of the delivery system. Recent research has revealed that arginine with guanidyl group can facilitate the cellular uptake of covalently conjugated particles, although the uptake mechanism is still controversial [37]. Wang et al. [37] synthesized rod-like arginine functionalized HA (HA/Arg) nanoparticles with an average length of 50–90 nm and the DNA complex was prepared by mixing the HAp/Arg with plasmid DNA (pEGFP-N1). Transfection efficiency was tested on the HeLa cells which were treated with pEGFP-N1-HA/Arg complex and the transfection efficiency increased after arginine modification.

In another study, it was reported that if two-dimensional layered materials are used as gene delivery materials, they exhibit desired functionalities [41]. Zuo et al. [41] prepared lamellar HA (L-HA) nanoplates with varying size and shape by changing the amount of the templating agent and precursor solutions. pDNA was added dropwise into L-HA suspensions with varying concentrations for the complex formation. It was suggested that pDNA loading is not only limited by the adsorption of the HA surface but also intercalation of pDNA into HA layers. Thus, a mouse fibroblast L929 cell line was used to investigate the gene transfection of pDNA/L-HA complexes. As the L-HA showed much better DNA loading capacity and transfection yield, the great potential of the nanoplates with good lamellar structure was reported.

In summary, different strategies were developed for improving the gene transfection efficiency by using CaP nanoparticles as non-viral vectors and some of the reported models are illustrated in **Figure 3**. Most of the approaches were focused on controlling the size and the shape by utilizing special agents during CaP synthesis or encapsulating the particle-DNA complex mostly in a proper layer of polymer to obtain positive surface charges.

3.2. The use of calcium phosphate nanoparticles for gene silencing

Small interfering RNA (siRNA) is a class of short, double-stranded RNA molecules of about 21–25 nucleotides, which can mobilize the RNA interference (RNAi) pathway [14, 42]. RNA

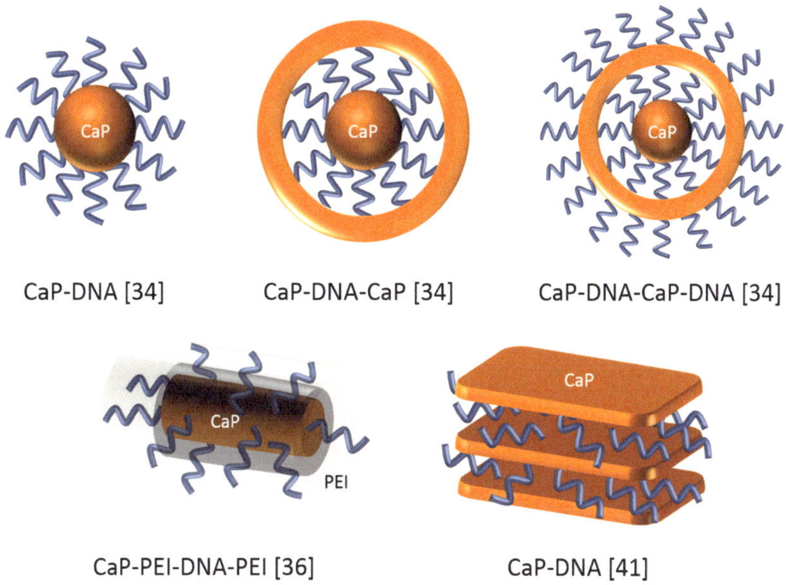

Figure 3. Some of the CaP-DNA complex strategies reported in the literature.

interference (RNAi) effect is a sequence-specific gene-silencing process which is effectively applied for functional genomics, in vivo target validation, and gene-specific medicines. Similar to the antisense and ribozyme strategies, RNAi depends on complementarity between the RNA and its target mRNA for the destruction of the target gene [42].

When compared to conventional drugs or small molecules with a limited range of protein targets, siRNA may perhaps be used to interfere with the expression of nearly any gene transcript in a specific manner. Although siRNA-based RNAi technology has been launched as an effective methodology in treating several genetic diseases, this technology is limited owing to the degradation of siRNA in the nucleases [14]. Thus, a biocompatible carrier approach is necessary to overcome the barriers for the delivery of siRNAs for successful therapeutic applications [14].

Although both plasmid DNA and siRNA are similar double-stranded nucleic acids having anionic phosphodiester backbones and the same negative charge to nucleotide ratio, the use of DNA requires some different major aspects regarding its complex formation and its intracellular delivery [39]. For being transcribed, plasmid DNA needs to cross the cytosol and reach the nucleus. Therefore, the delivery of plasmid DNA by using chemical delivery vectors is sometimes ineffective on non-dividing quiescent cells due to the cell nucleus envelope barrier [42]. Further, the siRNA only requires to be delivered to the cytosol for initiating the silencing activity and the therapeutic potential of siRNAs has prompted a revival and strong research and development effort in this field.

In some studies, CaP is precipitated with siRNA to form the gene silencing complex. Sokolova et al. [5] investigated the efficiency of single-shell (the core of CaP, the shell of siRNA) and multi-shell (the core of CaP, a first shell of siRNA, the second shell of CaP, the third and outermost shell of siRNA) nanoparticles to inhibit the expression of intercellular enhanced green fluorescent protein (EGFP) in HeLa cells. Both single-shell and multi-shell nanoparticles have a spherical shape with a diameter of about 100–200 nm. The gene silencing efficiency is proportional to the extent of EGFP inhibition in the cells (cells which show no green fluorescence anymore) and determined by transmission light microscopy and fluorescence microscopy. Not only the concentration of oligonucleotides but also the nature of oligonucleotides (single- or double-stranded) can affect the stability of the colloidal delivery system. Monodisperse colloids can be prepared with siRNA at a concentration of 45 mM. The relative gene silencing efficiencies of different delivery systems were investigated and high gene silencing efficiency was obtained for multi-shell nanoparticles in which siRNA is protected from intracellular degradation.

Although the use of CaP for siRNA delivery has performed a lot of promising progresses, the application of CaP-siRNA in clinical therapy is still far from satisfaction [31]. To improve the physical stability and the transfection efficiency of CaP nanoparticles, different strategies have been evaluated for the modification of CaP carriers. One of the most crucial factors of a successful siRNA therapy is to control the size and the surface charge of the siRNA complex for ensuring the delivery and release of the therapeutic material into the target gene [31]. For this aim, some surface modifications including encapsulating the CaP-siRNA complex with different polymers or adding a stable polymer outer layer were reported [43]. In order to obtain better colloidal stability and to reduce protein adsorption, Polyethylene glycol (PEG), a neutral and hydrophilic polymer, could be used [44]. It was reported that, compared with some commercially available transfection reagents, CaP-PEG nanoparticles reveals better serum tolerability. Owing to the decomposition of CaP nanoparticles under a cytosolic environment, calcium ion concentration increases. However, by using PEG as a shell on the CaP core, much lower calcium ion concentrations can be maintained and thus significant biological effects can be reached in cultured cell lines. Giger et al. [44] modified CaP nanoparticles with PEG-alendronate (alendronate is a bisphosphonate used to treat osteoporosis) for siRNA delivery and it was reported that PEG-ALE could stabilize the nanoparticles better. The transfection efficiency of the nanoparticles was assessed with a model siRNA targeting the Bcl-2 protein which inhibits apoptosis and is overexpressed in several cancers. PEG-ALE-CaP-siRNA exhibited a strong silencing effect in vitro at both the mRNA and protein levels. The cellular trafficking study showed that PEG-ALE-CaP-siRNA internalized into cells relied largely on the clathrin-dependent endocytosis. They further used PEG-functionalized bisphosphonate (PEG-bp) to prepare bp-stabilized CaP nanoparticles with the size of 200 nm for gene delivery [45]. PEG-bp-CaP showed effectively and sustained transfection ability to cells in vitro with low toxicity. Zhang et al. [18] developed CaP nanomicellar carriers which can efficiently keep siRNA in their CaP inner core by using PEG for regulating the size of the nanoparticles, which facilitated the internalization of siRNA by cells. The nanoparticles exhibited a well-defined spherical shape with a diameter of 100 nm and the thickness of PEG shell is about 10 nm. To evaluate the gene-silencing efficiency, dual luciferase assay was applied to the nanoparticles having different PEG shell thicknesses. Compared to 15 kDa PEG, the samples prepared with 12-kDa PEG showed better gene silencing efficiency. So, the difference in gene knockdown may stem from the difference in cellular uptake of the nanoparticles with different PEG shell thicknesses. Further, compared to the samples prepared at pH 8.0, the samples prepared at pH 7.5 showed higher gene knockdown levels under the same conditions.

Polyethyleneimine (PEI), another hydrophilic polymer, can also be used as an outer layer of the CaP-siRNA complex. However, it should be noted the potential cytotoxicity may limit its gene or drug delivery-based clinical applications [35]. Although PEI is an efficient transfection agent for DNA due to its great endosomal escaping capability, PEI/siRNA complexes are highly unstable [46]. Nevertheless, some successfully in vitro siRNA delivery applications were reported by using PEI as a surface modifier. Klesing et al. [36] developed a cationic nucleic acid delivery system based on hydroxyapatite nanorods functionalized with a layer of PEI. For gene silencing, HeLa-EGFP cell line (a genetically modified transformed cervix epithelial cell line expressing the enhanced green fluorescent protein, EGFP) was used. Commercial carriers, Lipofectamine™ and PolyFect®, were also used for comparing the transfection efficiency of the CaP-PEI system. Gene silencing efficiency of hydroxyapatite–PEI–siRNA nanoparticles on HeLa-EGFP cells was determined approximately 48–72 h after transfection by transmission light microscopy and fluorescence microscopy. The gene silencing efficiency of HeLa-EGFP cells was systematically determined for increasing amounts of added siRNA, up to a charge reversal by the anionic siRNA. The efficiency of gene silencing increased with higher amounts of siRNA, however, minimized at the point where particles agglomerated. By adding a PEI layer around the hydroxyapatite/PEI/siRNA, the charge reversed back to positive, resulting in even higher silencing rates.

To enhance the silencing efficiency, Neuhaus et al. [47] generated triple-shell CaP nanoparticles (CaP–siRNA–CaP–PEI) with an outer shell of PEI. PEI was used for the colloidal stabilization and to give the particles a positive charge. Tumor necrosis factor α (TNF-α), which is a pro-inflammatory cytokine, was silenced by treating LPS-stimulated MODE-K cells with triple-shell CaP nanoparticles. MODE-K cells are a murine epithelial cell line from the intestine and serve as a model for intestinal epithelial cell function. In this study, the gene silencing efficiency was first demonstrated with HeLa–EGFP cells by three different methods: cell counting by using fluorescence microscopy, fluorescence-activated cell sorting (FACS) analysis, and quantitative real-time PCR (qPCR). The silencing efficiency by fluorescence microscopy was difficult to interpret. More accurate results were obtained by FACS analysis and qPCR. qPCR results show enhanced silencing effects with the increasing amount of siRNA. By introducing triple-shell CaP nanoparticles into MODE-K cells, the expression of TNF-α was strongly downregulated. Gene silencing efficiencies close to results with the commercial agent Lipofectamine were achieved. Thus, CaP nanoparticles are well suited to reduce inflammatory reactions by silencing the corresponding cytokines, e.g., TNF-α.

In most of the studies, siRNA was precipitated together with calcium and phosphor precursor solutions for obtaining core-shell delivery complexes. Therefore, the particles were only characterized in terms of particle size and surface charges, while it was not determined if the particles were hydroxyapatite or other CaP phases. However, according to the current state of knowledge; particle size, particle shape, surface charge, and crystallinity may have an effect on the siRNA binding capacity and thus the efficiency of cellular uptake. For this reason, Bakan et al. [14] synthesized spherical-like hydroxyapatite (HA-s), needle-like hydroxyapatite (HA-n), and calcium deficient hydroxyapatite (CDHA), and the particle characteristics were investigated in detail. The effects of particle size, particle shape, crystallinity, and surface charge on the binding capacity of two different siRNAs were investigated. Arginine functionalized needle-like HA nanoparticles with a diameter of 15 nm and a length of 100–150 nm provided the best binding behavior for siRNA among the others, due to the high positive zeta potential and the geometry factor of the particles. Needle-like HA nanoparticles have also a much larger surface to volume ratio, hence, the surface area of the Ca^{2+} rich plane, which is considered to be positively charged, is higher than the others. Therefore, the binding efficiency of HA-n is higher than HA-s. Although CDHA has similar geometry to HA-n with almost the same dimensions, the binding potential of CDHA is lower than HA-n due to the deficiency of positive charged Ca^{2+}. Additionally, the nanoparticles without any functionalization also exhibited considerable affinity to the siRNAs.

3.3. The use of calcium phosphate nanoparticles for in vivo gene silencing

The success of in vitro siRNA therapy has been demonstrated by inhibiting the expression of many different genes and in many different cell types [48]. However, siRNA delivery to animal tissues requires considering some physical, chemical, and biological complicated processes [48]. There are some important weaknesses that limit the extensive application of several delivery systems such as the mutagenesis or oncogenesis potential, host immune responses, and high cost. As the immune response toward the non-viral siRNA delivery systems is very few, this yields an advantage in drug target validation and permits the therapeutic applications of siRNA [49].

$CaCl_2$-modified HA nanoparticles were used as a non-viral vector in the in vivo delivery of si-Stat3 plasmid-based siRNA for suppressing the mouse prostate tumor growth and promoting apoptosis of cancer cells [49]. Stat3 is a transcription factor and has been found in an activated state in numerous primary tumors and the Stat3 signaling pathway can be used as a potential target for antitumor therapy. The levels of Stat3 mRNA and protein were downregulated in cancer tissues after si-Stat3 treatment, indicating that HA can deliver Stat3-specific siRNA into cancer cells, resulting in the inhibition of Stat3 expression.

The effect of HA-siRNA complex on chronic inflammatory pain was investigated in vivo in mice [50]. HA nanorods with a length of 40–50 nm were synthesized by a chemical precipitation-hydrothermal technique for the delivery of siRNA. NR2B, the target gene, is a NMDA receptor subunit which plays an important role in the adjustment of chronic pain. To obtain HA-NR2B-siRNA complex, HA nanoparticle suspension was mixed with siRNA under different pH conditions. The results show that HA nanoparticles can bind to siRNA effectively in the acidic or neutral condition at an HA: siRNA mass ratio of 35:1 or higher. In addition, HA-NR2B-siRNA complex remains stable in the physiological condition. HA-siRNA complex was injected into subarachnoid space of each mice. On the seventh day after injection, 1% formalin solution was injected into the plantar surface of the right hind paw for the formalin test. The contralateral paw served as control. Pain scores were given and the Time-Score Curves were plotted. Via intrathecal injection, HA-siRNA complex can significantly reduce formalin-induced nociception in the tonic phase in mice. Therefore, HA may have a potential as an effective siRNA carrier even in in vivo.

In another study, multi-shell CaP nanoparticles (CaP-siRNA) encapsulated into poly(D,L-lactide-co-glycolide) (PLGA) and a final layer of polyethyleneimine (PEI) for the local therapeutic treatment of colonic inflammation [51]. The synthesis of nanoparticles was performed via rapid mixing of equal amounts of calcium-L-lactate and diammonium hydrogen phosphate aqueous solutions. Right after mixing, the CaP dispersion was mixed with siRNA and then encapsulated into PLGA. The freeze-dried nanoparticles were afterward resuspended in an aqueous solution of PEI. PLGA allows for a controllable releasing behavior which can be affected by the geometry and the molecular weight of the nanoparticles [51]. PEI was employed as a shell layer to provide a positive surface charge to the encapsulated nanoparticles. siRNA-loaded nanoparticles could be effectively delivered into the cytoplasm of epithelial cells and immune cells of mice in vivo and thus induce active gene silencing. Such a treatment induced a significant reduction of the target genes (TNF-α, IP-10, and KC) in colonic biopsies and mesenteric lymph nodes when they were accompanied with a distinct amelioration of intestinal inflammation. Thus, this study provides a promising approach for the treatment of intestinal inflammation via CaP/PLGA-siRNA nanoparticles.

3.4. Calcium phosphate nanoparticles for miRNA delivery

MicroRNAs (miRNAs; miRs) are single-stranded, non-coding RNA molecules which are biochemically and functionally indistinguishable from siRNAs [52]. It is known that miRNA can bind to the target mRNA and inhibits its expression [52]. Although the functions of miRNAs have not quite known yet, the role of miRNA in the regulation of gene expression by controlling the various cellular and metabolic pathways was revealed. When a single miRNA is able to regulate several mRNA targets, multiple miRNAs are cooperatively able to regulate a single mRNA target [53].

As similar to siRNA applications, the major challenge of miRNA therapy is delivering the miRNA to the target tissue efficiently [54]. The nucleases (e.g., serum RNase A-type nucleases in the blood) can degrade naked miRNAs within seconds, therefore, employing a sophisticated carrier is crucial to enhance the intracellular delivery and provide an endosomal escape [54]. Current non-viral technologies used Lipofectamine®2000, PEG or PEI as a carrier for the protecting of miRNAs [55]. However, CaPs are a promising alternative to polymers or synthetic lipids with having better biocompatibility and the ease of use [55]. Manipulation of gene expression by using microRNAs (miRNAs) offers remarkable and promising potential for the field of tissue engineering [55]. Nevertheless, the deficiency of adequate site-specific and bioactive carriers has limited the clinical translation of miRNA-based therapies. For this purpose, Castano et al. [55] developed a novel non-viral delivery complex by combining HA nanoparticles with reporter miRNAs (nanomiRs). To investigate the transfection efficiency of this complex, human mesenchymal stem cells (hMSCs) which are defined as a particularly hard cell type to efficient transfection was used. The results demonstrate that HA nanoparticles combined with miR-mimics and antagomiRs resulted in high cellular internalization in monolayer hMSCs with limited cytotoxicity and both nanomiR-mimics and nanoantagomiRs provided continuous interference of greater than 90% in monolayer over 7 days. Although the efficient intracellular internalization generally requires positively charged particles of less than 200 nm, negatively charged HA-miRNA complexes with approximately 300 nm size provided high cellular internalization in accordance with the related literature [55].

Mauro et al. [56] investigated the potential of negatively charged CaP nanoparticles as a therapeutic system for the intracellular delivery of miRNAs to the cardiac tissue. In this study, CaP nanoparticles were synthesized using citrate as a stabilizing agent and crystal growth regulator. During the CaP synthesis, different amounts of synthetic unmodified and unprotected miRNA duplexes (ranging from 1 to 10 $\mu g\ ml^{-1}$) were added to the reaction medium. Although the size of CaP-miRNAs increased with the increasing amount of miRNA, the PDI values and surface charges stayed close to those of miRNA-free nanoparticles. To evaluate the efficiency of miRNA delivery, HL-1 cells were exposed to CaP-miRNAs both in vivo and in vitro and the intracellular levels of miRNAs were measured at different times after the transaction. The results show that CaP nanoparticles can be successfully penetrated into cardiomyocytes without increasing toxicity or mediating with any functional properties and synthetic microRNAs can be successfully encapsulated within CaP nanoparticles.

In summary, it was revealed that both positively and negatively charged HA nanoparticles have the potential to complex with both miR-mimics and antagomiRs for forming nanomiRs, and they have a tremendous capability to maintain a successful internalization. Therefore, HA nanoparticles can be employed as a delivery system for the clinical translation of miRNA-based therapies.

4. Conclusion

The yield of gene therapy generally relies on the performance of the carrier vector. Because of the high efficiency of viral vectors in gene therapy, their use has been prevailing so far in the clinical trials. However, their potential immunogenicity, cytotoxicity, and mutagenesis risks are the major drawbacks of using them. Therefore, the use of a variety of non-viral vectors entering clinical trials is because of the advances in their efficiency, specificity, gene expression duration, and safety characteristics. In order to design bio-adaptable non-viral vectors, mechanisms of transfection, from complex formation to intracellular delivery, have to be well-understood. In the gene delivery process through of non-viral vectors, there have been potential rate-limiting stages such as, cell membrane interaction efficiency, internalization, endosomal escape, gene release from intracellular compartments and the transcription into the nucleus. Each of the aforementioned stages has to be determined by examining the characteristics at the molecular level for coming through the sorted limitations. Clinical achievement is still beyond the ideal conditions despite the fact that non-viral vectors can be applied efficiently for in vitro trials. Promisingly, advances in novel technologies for the development of these carriers allow many research groups for studying in vitro and in vivo behaviors of non-viral vectors for gene therapy.

Calcium phosphates (CaPs) are the main mineral constituents of biological hard tissues. Except for enamel, they are always found as nanoparticles. Due to their excellent features, they have widespread use as a bionanomaterial in the fields of nanomedicine, including tissue engineering, imaging, and drug/gene delivery. They are biocompatible, biodegradable, non-toxic, and non-immunogenic. Moreover, its characteristic degradation into ions is present in all cells and extracellular fluids make it a safe and suitable intracellular delivery system for therapeutic small molecules and genes. CaP nanoparticles have been widely used in non-viral gene delivery systems since the 1970s. Negatively charged nucleic acids can bind to CaP nanoparticles with high affinity by calcium ion chelation and are consequently protected from degradation by endonucleases. During internalization, continuous acidification in the lysosomes causes degradation of CaP particles and the therapeutic molecules can be released. Nevertheless, the cytotoxicity may slightly increase owing to the rise in the level of intracellular calcium which is caused by this degradation. In the literature, it has been reported that a large spectrum of CaP nanoparticles has been synthesized and employed as a carrier for DNA, siRNA, and miRNA. In most of the studies, CaP-nucleic acid complex was obtained by quick mixing of calcium precursor solution with DNA/siRNA which is followed by adding the phosphate solution. Since the nucleation and crystal growth can be influenced by the precipitation parameters such as pH, concentrations of each solution and temperature and also, cell transfection efficiency may be drastically affected by similar conditions. The time slot between precipitation and transfection also influences the transfection efficiency in terms of colloidal stability of the complex. The efficiency of transfection has been found to be cell-type dependent and intercellular trafficking pathways. If DNA/siRNA is surrounded by an additional shell of calcium phosphate, it can be better protected from enzymatic degradation and intracellular attack by nucleases. Core-shell structure leads to a significant enhancement in the transfection or silencing efficiencies compared to single-shell calcium phosphate nanoparticles. It should also be noted that the surface charge of the complex has a significant effect on the penetration through the cell membrane. Due to the surface-exposed phospholipids, the cell membrane is negatively charged and positively charged particles are taken up by the membrane at a higher rate compared to negatively charged ones. For having positively charged CaPs, surface modification strategies including amino acid functionalization, surface coating with cationic or neutral polymers were carried out. Recent research has indicated that the modification of CaP nanoparticles by mostly coating a suitable polymer, provides a better colloidal stability and thus improves not only in vitro but also in vivo gene transfection or silencing efficacy compared to using bare CaP nanoparticles. The amount of the therapeutic materials largely depends on the carrier adsorption potential and this drastically affects the surface charge of the complex; therefore, the higher the amount of DNA or siRNA does not bring about the higher transfection or silencing efficiency. To have a better understanding of the effect of carrier shape on cellular uptake and cytotoxicity, various studies have been carried out and reported in the literature. It was found that the cytotoxicity of the particles not only depends on the particle shape but also it can be cell-type dependent and the mechanisms that induce cytotoxicity through particle-cell association, whether arising from ROS generation, cytokine production or cellular uptake, are still not clear.

The effects of chemical, physical, and/or biological aspects generally make in vivo DNA or siRNA delivery complicated for animal tissues. The limitations in the use of most delivery systems in vivo can be summarized as responses of host immune, the risk of mutagenesis or oncogenesis and high costs. Considering CaP nanoparticles as non-viral vectors are still to be questioned for clinical trials. The increase in the level of the solution's ion strength, for instance, might cause a decrease in the electrostatic repulsion between nanoparticles and this might result in agglomeration. In the presence of proteins, hydrophilicity, surface chemistry, and the charge of the nanoparticles can be altered due to adsorption of proteins on their surface. Accordingly, the behavior of a functionalized nanoparticle will be different in the tissue or the bloodstream compared to aqueous medium or physiologic solutions. This case leads to complications which are difficult to handle in the clinical trials. Nanoparticles have completely different biological properties when compared to their bulk properties and therefore all aspects are to be considered carefully for maintaining appropriate colloid stability and cellular interaction behavior. In the modified nanoparticles, the physicochemical state becomes more complex and this leads to less predictable biological responses and the cytotoxicity of the modified systems are also to be considered.

Bringing together all outcomes of the recent research, it can be concluded that the use of CaPs as non-viral vectors prevails among other non-viral vectors owing to their low-cost and safety behavior, as well as they exhibit better performance in overcoming the limitations that have been observed in other vectors. Providing that the size of the delivery complex is small enough, the surface charge of the complex permits maximum cellular internalization and the therapeutic molecules are protected through the cellular uptake; a successful transfection can be achieved and CaP nanoparticles can be defined as "second generation non-viral vectors" as Maitra [23] denoted.

Acknowledgements

I would like to cordially thank Dr. Meltem Sezen, Dr. Turgay Han, and Göknur Kara for their valuable contribution to this book chapter. Also support by Dr. Volkan Özgüz, the director of Nanotechnology Research and Application Center at Sabanci University, is gratefully acknowledged.

Author details

Feray Bakan

Address all correspondence to: feraybakan@sabanciuniv.edu

Sabanci University Nanotechnology Research and Application Center (SUNUM), Istanbul, Turkey

References

[1] Epple M, Ganesan K, Heumann R, Klesing J, Kovtun A, Neumann S, Sokolova V. Application of calcium phosphate nanoparticles in biomedicine. Journal of Materials Chemistry. 2010; **20**(1):18-23

[2] Bisht S, Bhakta G, Mitra S, Maitra A. pDNA loaded calcium phosphate nanoparticles: Highly efficient non-viral vector for gene delivery. International Journal of Pharmaceutics. 2005;**288**(1):157-168

[3] Rayburn ER, Zhang R. Antisense, RNAi, and gene silencing strategies for therapy: Mission possible or impossible? Drug Discovery Today. 2008;**13**(11):513-521

[4] Chirila TV, Rakoczy PE, Garrett KL, Lou X, Constable IJ. The use of synthetic polymers for delivery of therapeutic antisense oligodeoxynucleotides. Biomaterials. 2002;**23**(2):321-342

[5] Sokolova V, Kovtun A, Prymak O, Meyer-Zaika W, Kubareva EA, Romanova EA, Oretskaya TS, Heumann R, Epple M. Functionalisation of calcium phosphate nanoparticles by oligonucleotides and their application for gene silencing. Journal of Materials Chemistry. 2007;**17**(8):721-727

[6] Zhang Y, Satterlee A, Huang L. In vivo gene delivery by nonviral vectors: Overcoming hurdles? Molecular Therapy. 2012;**20**(7):1298-1304

[7] Gascón AR, del Pozo-Rodríguez A, Solinís MA. Non-Viral Delivery Systems in Gene Therapy, Gene Therapy - Tools and Potential Applications, Dr. Francisco Martin (Ed.), InTech: Rijeka, Croatia. 2013. DOI: 10.5772/52704

[8] Ramamoorth M, Narvekar A. Non-viral vectors in gene therapy—An overview. Journal of Clinical and Diagnostic Research. 2015;**9**(1):1-6

[9] Robbins PD, Ghivizzania SC. Viral vectors for gene therapy. Pharmacology and Therapeutics. 1998;**80**(1):35-47

[10] Sokolova V, Epple M. Inorganic nanoparticles as carriers of nucleic acids into cells. Angewandte Chemie International Edition. 2008;**47**(8):1382-1395

[11] Yin H, Kanasty RL, Eltoukhy AA, Vegas AJ, Dorkin JR, Anderson DG. Non-viral vectors for gene-based therapy. Nature Reviews Genetics. 2014;**15**(8):541-555

[12] Dorozhkin SV, Epple M. Biological and medical significance of calcium phosphates. Angewandte Chemie International Edition. 2002;**41**(17):3130-3146

[13] Okazaki M, Yoshida Y, Yamaguchi S, Kaneno M, Elliott JC. Affinity binding phenomena of DNA onto apatite crystals. Biomaterials. 2001;**22**(18):2459-2464

[14] Bakan F, Kara G, Cokol Cakmak M, Cokol M, Denkbas EB. Synthesis and characterization of amino acid-functionalized calcium phosphate nanoparticles for siRNA delivery. Colloids and Surfaces B: Biointerfaces. 2017;**158**:175-181

[15] Khan MA, VM W, Ghosh S, Uskokovic V. Gene delivery using calcium phosphate nanoparticles: Optimization of the transfection process and the effects of citrate and poly (L-lysine) as additives. Journal of Colloid and Interface Science. 2016;**471**:48-58

[16] Zhang J, Sun X, Shao R, Liang W, Gao J, Chen J. Polycation liposomes combined with calcium phosphate nanoparticles as a non-viral carrier for siRNA delivery. Journal of Drug Delivery Science and Technology. 2015;**30**:1-6

[17] Turkez H, Yousef MI, Sonmez E, Togar B, Bakan F, Sozio P, Stefano AD. Evaluation of cytotoxic, oxidative stress and genotoxic responses of hydroxyapatite nanoparticles on human blood cells. Journal of Applied Toxicology. 2014;**34**(4):373-379

[18] Zhang M, Ishii A, Nishiyama N, Matsumoto S, Ishii T, Yamasaki Y, Kataoka K. PEGylated calcium phosphate nanocomposites as smart environment-sensitive carriers for siRNA delivery. Advanced Materials. 2009;**21**(34):3520-3525

[19] Freitag R, Vogt S, Modler M. Thermoreactive displacers for anion exchange and hydroxyapatite displacement chromatography. Biotechnology Progress. 1999;**15**(3):573-576

[20] Bakan F, Lacin O, Sarac H. A novel low temperature sol–gel synthesis process for thermally stable nano crystalline hydroxyapatite. Powder Technology. 2013;**233**:295-302

[21] Girija EK, Kumar GS, Thamizhavel A, Yokogawa Y, Kalkura SN. Role of material processing on the thermal stability and sinterability of nanocrystalline hydroxyapatite. Powder Technology. 2012;**225**:190-195

[22] Wolff JA, Malone RW, Williams P, Chong W, Acsadi G, Jani A, Felgner PL. Direct gene transfer into mouse muscle in vivo. Science. 1990;**247**(4949 Pt 1):1465-1468

[23] Maitra A. Calcium phosphate nanoparticles: Second-generation nonviral vectors in gene therapy. Expert Review of Molecular Diagnostics. 2005;**5**(6):893-905

[24] Khalil IA, Kogure K, Akita H, Harashima H. Uptake pathways and subsequent intracellular trafficking in nonviral gene delivery. Pharmacological Reviews. 2006;**58**(1):32-45

[25] Yameen B, Choi WI, Vilos C, Swami A, Shi J, Farokhzad OC. Insight into nanoparticle cellular uptake and intracellular targeting. Journal of Controlled Release. 2014;**190**:485-499

[26] Graham FL, van der Eb AJ. A new technique for the assay of infectivity of human adenovirus 5 DNA. Virology. 1973;**52**(2):456-467

[27] Welzel T, Radtke I, Meyer-Zaika W, Heumann R, Epple M. Transfection of cells with custom-made calcium phosphate nanoparticles coated with DNA. Journal of Materials Chemistry. 2004;**14**(14):2213-2217

[28] Kakizawa Y, Miyata K, Furukawa S, Kataoka K. Size-controlled formation of a calcium phosphate-based organic-inorganic hybrid vector for gene delivery using poly(ethylene glycol)-block-poly(aspartic acid). Advanced Materials. 2004;**16**(8):699-702

[29] Wu GJ, Zhou LZ, Wang KW, Chen F, Sun Y, Duan YR, Zhu YJ, Gu HC. Hydroxylapatite nanorods: An efficient and promising carrier for gene transfection. Journal of Colloid and Interface Science. 2010;**345**(2):427-432

[30] Zhu SH, Huang BY, Zhou KC, Huang SP, Liu F, Li YM, Xue ZG, Long ZG. Hydroxyapatite nanoparticles as a novel gene carrier. Journal of Nanoparticle Research. 2004;**6**(2):307-311

[31] Olton D, Li J, Wilson ME, Rogers T, Close J, Huang L, Kumta PN, Sfeir C. Nanostructured calcium phosphates (nanoCaPs) for non-viral gene delivery: influence of the synthesis parameters on transfection efficiency. Biomaterials. 2007;**28**(6):1267-1279

[32] Uskokovic V, Uskokovic D. Nanosized hydroxyapatite and other calcium phosphates: Chemistry of formation and application as drug and gene delivery agents. Journal of Biomedical Materials Research B: Applied Biomaterials. 2011;**96 B**(1):152-191

[33] Chowdhury EH, Kunou M, Nagaoka M, Kundu AK, Hoshiba T, Akaike T. High-efficiency gene delivery for expression in mammalian cells by nanoprecipitates of Ca–Mg phosphate. Gene. 2004;**341**:77-82

[34] Sokolova VV, Radtke I, Heumann R, Epple M. Effective transfection of cells with multi-shell calcium phosphate-DNA nanoparticles. Biomaterials. 2006;**27**(16):3147-3153

[35] Xu X, Li Z, Zhao X, Keen L, Kong X. Calcium phosphate nanoparticles-based systems for siRNA delivery. Regenerative Biomaterials. 2016;**3**(3):187-195

[36] Klesing J, Chernousova S, Epple M. Freeze-dried cationic calcium phosphate nanorods as versatile carriers of nucleic acids (DNA, siRNA). Journal of Materials Chemistry. 2012;**22**(1):199-204

[37] Wang GH, Zhao YZ, Juan TAN, Zhu SH, Zhou KC. Arginine functionalized hydroxy-apatite nanoparticles and its bioactivity for gene delivery. Transactions of Nonferrous Metals Society of China. 2015;**25**(2):490-496

[38] Gratton SE, Ropp PA, Pohlhaus PD, Luft JC, Madden VJ, Napier ME, DeSimone JM. The effect of particle design on cellular internalization pathways. Proceedings of the National Academy of Sciences. 2008;**105**(33):11613-11618

[39] Rabinovich-Guilatt L, Couvreur P, Lambert G, Dubernet C. Cationic vectors in ocular drug delivery. Journal of Drug Targeting. 2004;**12**(9-10):623-633

[40] Vasir JK, Labhasetwar V. Quantification of the force of nanoparticle–cell membrane interactions and its influence on intracellular trafficking of nanoparticles. Biomaterials. 2008;**29**(31):4244-4252

[41] Zuo G, Wan Y, Meng X, Zhao Q, Ren K, Jia S, Wang J. Synthesis and characterization of a lamellar hydroxyapatite/DNA nanohybrid. Materials Chemistry and Physics. 2011;**126**(3): 470-475

[42] Scherman D, Rousseau A, Bigey P, Escriou V. Genetic pharmacology: Progresses in siRNA delivery and therapeutic applications. Gene Therapy. 2017;**24**(3):151-156

[43] Escriou V, Carrière M, Bussone F, Wils P, Scherman D. Critical assessment of the nuclear import of plasmid during cationic lipid-mediated gene transfer. The Journal of Gene Medicine. 2001;**3**(2):179-187

[44] Giger EV, Castagner B, Räikkönen J, Mönkkönen J, Leroux JC. siRNA transfection with calcium phosphate nanoparticles stabilized with PEGylated chelators. Advanced Healthcare Materials. 2013;**2**(1):134-144

[45] Giger EV, Puigmartí-Luis J, Schlatter R, Castagner B, Dittrich PS, Leroux JC. Gene delivery with bisphosphonate-stabilized calcium phosphate nanoparticles. Journal of Controlled Release. 2011;**150**(1):87-93

[46] Devarasu T, Saad R, Ouadi A, Frisch B, Robinet E, Laquerrière P, Voegel JC, Baumert T, Ogier J, Meyer F. Potent calcium phosphate nanoparticle surface coating for in vitro and in vivo siRNA delivery: A step toward multifunctional nanovectors. Journal of Materials Chemistry B. 2013;**1**(36):4692-4700

[47] Neuhaus B, Frede A, Westendorf AM, Epple M. Gene silencing of the pro-inflammatory cytokine TNF-α with siRNA delivered by calcium phosphate nanoparticles, quantified by different methods. Journal of Materials Chemistry B. 2015;**3**(36):7186-7193

[48] PY L, Xie F, Woodle MC. In vivo application of RNA interference: From functional genomics to therapeutics. Advances in Genetics. 2005;**54**:115-142

[49] Liang ZW, Guo BF, Li Y, Li XJ, Li X, Zhao LJ, Gao LF, Yu H, Zhao XJ, Zhang L, Yan BX. Plasmid-based Stat3 siRNA delivered by hydroxyapatite nanoparticles suppresses mouse prostate tumour growth in vivo. Asian Journal of Andrology. 2011;**13**(3):481-486

[50] Yang H, Huang D, Zhu SH, Yan XB, Gu YH, Hui YXB, Gu YH, Yan H, Wu LX. Preparation and application of hydroxyapatite (HA) nanoparticles/NR2B-siRNA complex. Transactions of Nonferrous Metals Society of China. 2008;**18**(4):913-918

[51] Frede A, Neuhaus B, Klopfleisch R, Walker C, Buer J, Müller W, Epple M, Westendorf AM. Colonic gene silencing using siRNA-loaded calcium phosphate/PLGA nanoparticles ameliorates intestinal inflammation in vivo. Journal of Controlled Release. 2016; **222**:86-96

[52] MacFarlane LA, Murphy PR. MicroRNA: Biogenesis, function and role in cancer. Current Genomics. 2010;**11**(7):537-561

[53] Bartel DP. MicroRNAs: Genomics, biogenesis, mechanism, and function. Cell. 2004;**116**(2): 281-297

[54] Chen Y, Gao DY, Huang L. In vivo delivery of MiRNAs for cancer therapy: Challenges and strategies. Advanced Drug Delivery Reviews. 2015;**81**:128-141

[55] Castaño IM, Curtin CM, Shaw G, Murphy JM, Duffy GP, O'brien FJ. A novel collagen-nanohydroxyapatite microRNA-activated scaffold for tissue engineering applications capable of efficient delivery of both miR-mimics and antagomiRs to human mesenchymal stem cells. Journal of Controlled Release. 2015;**200**:42-51

[56] Di Mauro V, Iafisco M, Salvarani N, Vacchiano M, Carullo P, Ramírez Rodríguez GB, Patrício T, Tampieri A, Miragoli M, Catalucci D. Bioinspired negatively charged calcium phosphate nanocarriers for cardiac delivery of microRNAs. Nanomedicine. 2016;**11**(8): 891-906